Robert Ultzmann, Walter B. Platt

Pyuria or Pus in the Urine, and Its Treatment

comprising the diagnosis and treatment of acute and chronic urethritis,

prostatis, cystitis, and pyelitis, with especial reference to their local

treatment

Robert Ultzmann, Walter B. Platt

Pyuria or Pus in the Urine, and Its Treatment
comprising the diagnosis and treatment of acute and chronic urethritis, prostatis, cystitis, and pyelitis, with especial reference to their local treatment

ISBN/EAN: 9783337734633

Printed in Europe, USA, Canada, Australia, Japan

Cover: Foto ©berggeist007 / pixelio.de

More available books at **www.hansebooks.com**

PYURIA;

OR,

PUS IN THE URINE, AND ITS TREATMENT.

COMPRISING THE

DIAGNOSIS AND TREATMENT OF ACUTE AND CHRONIC URETHRITIS, PROSTATIS, CYSTITIS, AND PYE-LITIS, WITH ESPECIAL REFERENCE TO THEIR LOCAL TREATMENT.

By Dr. ROBERT ULTZMANN,

PROFESSOR OF GENITO-URINARY DISEASES IN THE VIENNA POLYCLINIC.

TRANSLATED BY PERMISSION BY

Dr. WALTER B. PLATT, F. R. C. S. (Eng.),

DEMONSTRATOR OF SURGERY IN THE UNIVERSITY OF MARYLAND; VISITING SURGEON TO BAYVIEW HOSPITAL, BALTIMORE.

NEW YORK:

D. APPLETON AND COMPANY,

1, 3, AND 5 BOND STREET.

1884.

TRANSLATOR'S PREFACE.

No man need apologize for urging any improvement in the treatment of any disease, whether this latter be self-inflicted or not. The mischief is done when the patient presents himself to the surgeon, and it is the business of the latter to restore the unfortunate man to his normal condition as soon as possible. If kind Nature usually does this after a time, should a surgeon refuse his aid in hastening a cure? The same spirit that would formerly reject the use of anæsthetics in childbirth, because pain is the inheritance of sin, and to-day leaves the foundling to die from neglect and exposure, declines to aid one suffering from certain painful and disabling diseases. May this narrow and illiberal spirit speedily pass away!

Professor Ultzmann, now well known to many English-speaking surgeons, strongly insists upon the *local* treatment of most of the diseases discussed in this monograph—accessible diseases, local in their origin and course. His brilliant results justify his methods. At the same time, it is to be hoped that such as read may also see that the author does not forget those chronic cases in feeble or neurotic individuals which are quickest cured by such general measures as change of climate, diet, and surroundings, and letting the urethra and bladder entirely alone. The translator hopes the medical profession may agree with him in believing Professor Ultzmann's "Pyuria" a real addition to our knowledge of the diagnosis and treatment of genito-urinary diseases.

W. B. P.

165 Park Avenue,
    Baltimore, Md.

CONTENTS.

P Y U R I A.*

INTRODUCTION.

WHEN more or less pus is passed in urination, we call the condition pyuria. It is, of course, evident that when we use the term *pyuria* we make a diagnosis in a very general way only. Like so many other expressions much used in former times, it is now seldom employed to express a disease in itself. It is like hæmaturia, albuminuria, etc. With the progress of medical diagnosis, these general terms gradually vanish ; and thus pyuria may be applied to several diseases of the apparatus designed for secreting, containing, and carrying off the urine, which can be accurately diagnosticated, by careful instrumental examination of the patient, and especially by a microscopical and chemical investigation of the urine.

In the following pages we will give all those aids which can help us to a special diagnosis as to what portion of the urinary apparatus is secreting the pus. Before doing this, we will discuss pus in its general relations, as the most important constituent of pyuric urine, and at the same time the best means to determine its presence.

* Clinical lectures given in the Vienna Allgem. Policlinic, Winter Semester, 1882–'83.

PYURIA.

Pus.—Pus consists of cellular elements, the pus-corpuscles, which are suspended in a fluid, pus-serum. Urine containing pus must admit of the detection of both these constituents, the pus-corpuscles as well as the serum. The pus-corpuscles are determined usually by the microscope ; the pus-serum, on the contrary, by chemical means, on account of the presence of albumen. Since pus consists of cellular elements, it necessarily follows that urine containing pus will appear more or less *turbid,* and turbid just in proportion to the amount of pus. A urine which has just been passed, and which appears clear and transparent, can never contain pus. The color of a pus-urine corresponds to the amount of coloring matter in the urine at the time, sometimes a lighter, again a darker wine-yellow ; pus itself has a greenish-yellow color ; this is, however, only imparted to the urine when it is present in large amount. The natural color of the pus is evident when it has separated from the urine as a compact sediment. Pus has an alkaline reaction, but the quantity of alkali is usually not enough to exert any considerable influence on the reaction of the urine. The acid reaction of the urine can only be neutralized or overcome by the alkali of the pus-serum when pus is present in a very great amount.

Pus-Corpuscles. — Pus-corpuscles are identical in shape and appearance with mucus-corpuscles, as well as with lymph and white blood corpuscles. Consequently, we can not always say, after simply examining a urinary sediment with the microscope, whether a given urine contains pus or only increased catarrhal secretion. Here the chemical examination puts us right. If the urine contain at the same time an amount of albumen corresponding to the quantity of corpuscles, and if these latter can have no other source than the pus-serum mixed with the urine, we infer the presence of *pus*. If, on the contrary, the urine contain no albumen, increased ca-

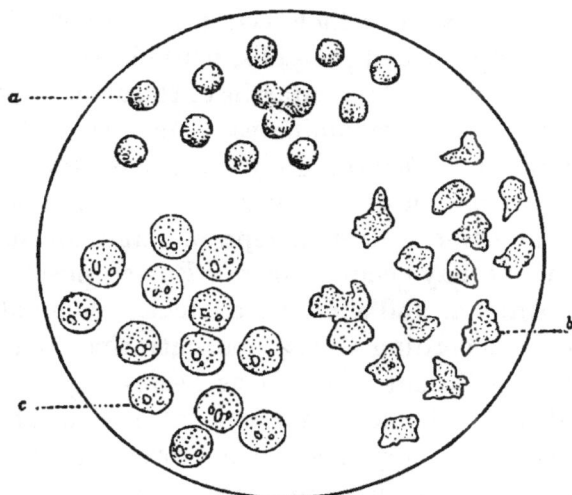

Fig. 1.—*a*. The usual globular pus-corpuscles of acid urine. *b*. Irregular pus-corpuscles of acid urine, provided with processes. *c*. Swollen pus-corpuscles of a strongly dilute or alkaline urine, with visible nuclei.

tarrhal secretion only is present. Pus-corpuscles always appear colorless under the microscope. They are somewhat larger than the red corpuscles, they have a diameter of $\frac{1}{80}$ to $\frac{1}{100}$ micro-millimetre, and they seem to be heavier than the latter, since, if we allow a sample of urine containing both blood and pus corpuscles intact to settle, the pus will form the lowest layer, while above, resting upon the green-yellow layer of pus, will be seen a bright red covering of red corpuscles. Pus-corpuscles are seen in two forms in acid urine : they are either globular or else irregular, sending out processes. The first is the variety most frequently met with. *Vogel* called attention to this difference in pus-corpuscles in the urine, and declared that their occurrence in the irregular form afforded a much less favorable prognosis than when they are globular. As a matter of fact, in those obstinate forms of pyuria which last for years, we find, as a rule, variously formed pus-cells, while in the milder and more transitory purulent catarrhs of the urinary passages, the globular form predominates. As a rule, the microscope shows us *either* the globular or else the polymorphous pus-corpuscles in a given urine, rarely both forms in the same urine, yet in exceptional and unusual cases both varieties may occur at once. Pus-corpuscles appear slightly granular and show no nucleus. They alter in a given urine according to the concentration and amount of saline constituents, as well as according to the reaction to litmus. In acid and concentrated urine, rich in salts, the pus-cells appear small and granular ; in alkaline urine, on the contrary, or in one of low specific gravity, they appear large and swollen ; the granulation of the protoplasm has vanished and the nuclei are plainly visible. In a very watery urine, the pus-corpuscles may appear two or three times as large as in a urine of normal concentration. This is the well-known change which pus-cells undergo when treated with distilled water. The alkalies, especially the carbonate of ammonia, of a urine undergoing alkaline fermentation, cause the pus-cells to swell still more, and they finally break up into a mass in which we no longer see the outline of the cells, but only the nuclei thus set free. A solution of iodine in potassic iodide colors the pus-corpuscles a fine yellow, while the nuclei which are now seen appear darker, and of a brownish-yellow color.

Pus-Serum.—Pus-serum, the intercellular fluid of the pus-corpuscles, is an opaque, pale-yellow fluid, always having an alkaline reaction. The alkaline reaction of the pus-serum arises from carbonates and basic phosphates of the alkalies and alkaline earths. The chief constituent of pus-serum is, however, serum albumen. Serum albumen of pus is not different from that of blood-serum. Moreover, pus-serum contains some paraglobulin and an alkali-albuminate. Since it is evident from the preceding that a pus-containing urine must be more or less opaque, and also contain an amount of albumen proportional to the pus, in order to detect the presence of the pus we proceed as follows, by what is, on the whole, the best method : Fill a test-tube half full with the urine to be tested, and heat gradually the upper half of the column of fluid to boiling. An increase of the opacity in the portion so heated, as compared with the lower portion which has not been boiled (as seen against a black background), indicates the presence of pus, if this increased opacity remains after the addition of one or two drops of acetic acid. The opacities in urine are briefly compared in the following table. This is, to be sure, not infallible, but it will always be of good service to the practitioner, on account of its simplicity, and the ease with which the tests may be made.

By gradually heating a given urine to boiling, the opacity—

Vanishes.	Increases.	Remains unchanged, even after addition of acetic acid.
If due to acid urates.	If due either to *earthy phosphates* (phosphaturia) or to *pus-corpuscles.* Add one or two drops of acetic acid:	The dimming is caused by cloudy *catarrhal secretion,* or by *bacteria.*

The dimness vanishes.	The dimness remains.
Phosphaturia.	*Pyuria.*

If, on the contrary, the supposed pus, after the urine has settled, forms a yellowish-white precipitate in the glass, visible to the naked eye, we carefully pour off the fluid above it, and add a few drops of a concentrated solution of caustic potassa (one third of water), shaking the vessel meanwhile until a visible change takes place. Now, if the sediment consists of pus, and if it is in the fine flaky state, with acid reaction of the urine (as, for example, in pyelitis), the entire sediment on addition of the caustic potassa (or caustic soda) will become transparent, thick, jelly-like, or at least capable of being drawn out in threads.

The change in the consistence of the purulent sediment in this test is occasioned by the transformation of the albumen in the pus into a glutinous, semi-fluid alkali-albuminate (Donne's pus test). If, on the contrary, the white pulverulent sediment consists of earthy phosphates, it will remain unaltered after the addition of caustic potassa or caustic soda, and retain its fluid consistence (can still be poured out in drops). This transformation of the pus into a transparent, tenacious, semi-fluid mass takes place sometimes even in the bladder itself, as in purulent vesical catarrh with ammoniacal fermentation of the urine, and is here occasioned by the carbonate of ammonia which is present. Likewise, these greenish-yellow, tenacious masses, resembling the discharge of nasal catarrh, which cling to the sides of the glass, are not composed of vesical mucus, but of alkaline pus. If retention of urine exists, along with pyuria and ammoniacal fermentation, and if much pus be present, this will be gradually changed into a honey-like, homogeneous, tenacious fluid, which flows off slowly when the catheter is passed, and on cooling takes on a sirupy consistence. On examining this urine microscopically, the contour of the pus-corpuscles can no longer be seen distinctly. We find simply their free nuclei. If an ulcerative process is present at the same time with pyuria in the urinary apparatus, the pus becomes ichorous, as is generally the case in an ulcerating neoplasm of the bladder. The urine at the same time has a dirty, greenish-brown color, and a penetrating, disagreeable, stinking odor, sometimes resembling fæces, sometimes ammonium sulphide. Such a urine has at the same time a strongly alkaline reaction. These are the urines which turn silver catheters black when employed in relieving the patient. Such a urine contains both considerable albumen and much blood-pigment. In the sediment, however, neither pus nor blood cells are to be seen, nor any epithelial structures. All cells are destroyed, and under the microscope we find only considerable detritus, triple phosphate, and bacteria in large amount. The penetrating odor developed by such a urine, which is made much more intense, sometimes indeed unendurable, by adding a mineral acid, may arise partly from the decomposition of albuminous bodies in the bladder, but

also partly because in ulcerative vesical processes (like-
• wise in parenchymatous cystitis) a diffusion of intestinal
gases into the bladder takes place readily. A similar
occurrence often takes place in abscesses or inflammations in the neighborhood of the intestines, as in peri-
typhlitic abscesses, into the fluid within the sacs of in-
carcerated hernias, etc.

The pus which is passed with the urine may have a
twofold source. It may have been formed on the surface of the mucous membrane of the urinary tract, or,
on the other hand, from the parenchyma of certain portions of the same, and is then genuine *abscess-pus.* As the
latter, it may also originate from purulent exudations and
abscesses situated around the urinary tract, and which
have opened into it. Thus abscesses of the kidney or
prostate, purulent parametritic and pericystitic exudations and other abscesses, not infrequently empty into
the bladder, and thus form the chief constituent of a
purulent sediment in a given urine. The varying
amount of pus present in a urine in such cases is in
itself presumptive proof of an addition of abscess-pus to
the urine. However, certainty is only afforded by microscopic examination of the sediment. That is to say, if
the pus comes from the mucous membrane alone, we find
microscopically, in addition to numerous pus-corpuscles,
the epithelium of the inflamed mucous membrane. If,
however, the pus comes from an abscess cavity which
has opened into the urinary tract, the granular cells, or
the so-called inflammation cells, which are always easy
to recognize, are never absent. Pus-corpuscles, under
the microscope, appear sometimes as solitary cells, sometimes conglomerated, again they are polymorphous, co-
hering masses. As conglomerated masses they not seldom arise from the small ducts of the accessory glands

of the urinary tract, or from the papillary* layer of the kidney itself, and they then take on a cylindrical form. From the character of the epithelial cells imbedded in these cylindrical casts, for example, whether covered with renal epithelium or spermatozoa, we derive in certain cases further information as to the source of the pus.

Pus, as passed with the urine, may come from the most diverse portions of the urinary tract. We can determine the following regions of its origin by the manner of urination itself, or by the more exact microscopic and chemical examination of the urine :

1. Pus arising from any part of the urethra, from the meatus to the compressor urethræ.

2. Pus arising from the neck of the bladder or the prostatic part.

3. Pus arising from the bladder.

4. Pus arising from the pelvis of the kidney and the kidney itself.

* These must not be confounded with the "gonorrhœal threads" to be described further on, and which are merely urethral epithelium rolled up into cylinders visible to the naked eye.

I.

PUS ARISING FROM THE URETHRA IN THE PART CONTAINED BETWEEN THE MEATUS AND THE COMPRESSOR URETHRÆ.—URETHRITIS.

This form of suppuration comprises the varieties of urethritis *per se*. The characteristic of this suppuration consists in the fact that the purulent catarrhal secretion is passed out with the urine and also spontaneously escapes from the meatus in the intervals of urination, discoloring the linen more or less, according to the intensity of the secretion. Suppuration in the urethra in this region alone, as far as the compressor, is never accompanied by tenesmus, or an almost uncontrollable desire to urinate. The patients urinate about as often as normally, and only perceive (especially in acute cases) a severe smarting sensation while the urine is passing along the urethra. Moreover, in acute cases of urethritis the mouth of the urethra is usually swollen and reddened, while in chronic cases, on the contrary, such an alteration is seldom seen at the meatus. As long as the purulent secretion of the urethra is more profuse, we invariably find the pus escaping as a drop from the meatus. If, on the contrary, the secretion is scanty, as in chronic urethritis, we usually see no secretion at all by external inspection. In order to see the secretion in such cases, we press the urethra from behind forward. In many cases we obtain a drop in this way. If, on the contrary, the secretion is thicker, more tenacious, and clings to the affected parts of the urethra more firmly, it may not be possible to detect the presence of secretion in this way. In such cases nothing remains but to have the patient urinate. The stream of urine, if much of any be in the bladder at the time, issues with considerable force, and, by reason of its friction with the urethral walls, washes off the purulent secretion, the coating of the diseased portion, and thus carries it out of the penis. At the same time, this membranous coating will be rolled up into a little thread, the so-called gonorrhœal threads or fibres, and thus eliminated. That this secretion of the urethra, called the " gonorrhœa thread," or pus-fibre, does not exist as such ready formed in the urethra, is shown by endoscopic examination of the latter, by which we find the coating simply clinging to the affected part, never free in the form of a thread. Gonorrhœal or pus fibres are of constant occurrence in a urine. These are to be distinguished from small portions of semen or mucus, which are likewise not infrequently found in urine, by the fact that they appear much more compact, and, by reason of their greater specific gravity, sink quickly to the bottom of the glass when urine is passed, while constituents like mucus or semen float about in the urine for some time as a light transparent cloud, at first incline more toward the surface, and later sink to the bottom. If we take up one of these things with a pair of fine forceps and place it under the microscope, then we can always tell whether the same consists of mucus, semen, or of purulent secretion. There is no other way to determine the quality of these small floating clouds or threads than by the microscope. We can only say that the secretion floating about in the urine as compact threads probably represents a purulent secretion. It

was formerly generally believed that the pus or gonorrhœal threads occurred *as such* in the urethra, and conclusions were drawn (which were thought correct) as
to the part of the urethra in which they originated,
from their form, length, and thickness. If the fibres
were short and equally thread-like throughout, they were
supposed to represent a cast of a gland-duct; if they
were long, thick, and cylindrical, they were supposed to
arise from the lumen of the urethra itself; and if the
threads were short, lumpy, or ragged, or had fringe-like
appendages, they were thought to come from the pars
prostatica. Since we know to-day that these fibres are
a product of the stream of urine, this reasoning, of course,
falls to the ground. We can only say that larger and
thicker threads come from a greater, smaller and thinner
threads from a less affected portion of the urethral mucous membrane. Further, we can say that the thread-
like gonorrhœal fibres for the most part come from
the anterior urethra as far as the compressor urethræ
muscle, and the broad, lumpy, and ragged affairs more
often from the posterior part of the urethra, from the
prostatic portion. This way of reasoning is not always
free from error. It is only when, by microscopical examination, in addition to the pus-corpuscles, we also find
spermatozoa imbedded in the gonorrhœal fibres, that
we can say, with the greatest probability, these originate
from the most posterior portion of the urethra.

Seen through the microscope, the pus or gonorrhœal
fibre consists of a transparent cylindrical mass in which
are imbedded numerous pus-corpuscles and a few urethral epithelial cells. The more compact such a fibre
seems, the more pus-cells it contains. If, on the contrary, the thread-fibre is delicate and transparent, it consists mostly of urethral epithelium and the pus-cells are

in the minority. The more the epithelium predominates in such a thread, the nearer the process is to a cure. Thus, the microscopic examination of these fibres has a certain significance in this respect. The suppuration of the urethra, as far as the compressor urethræ, is also dis-

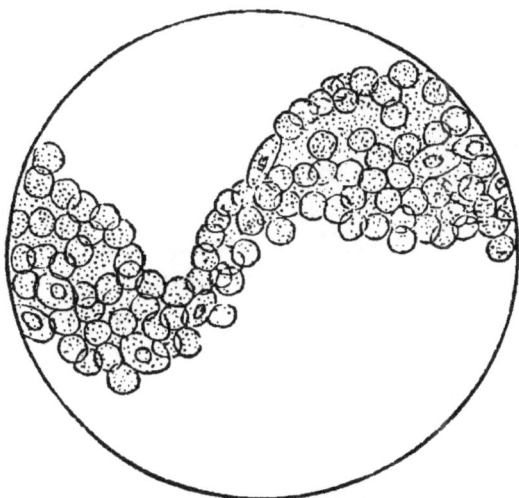

Fig. 2.—A so-called gonorrhœal thread, consisting of pus-corpuscles and urethral epithelium.

tinguished by the fact that when the patient urinates half his urine in one glass, the remainder in a second glass, only the first half will appear dim, while the second half of the urine will be found clear and transparent. That is to say, the urethra is freed from its secretion at the beginning of the urination, the urethra is completely washed out by the forcible stream, and consequently the secretion of the urethra is found in the first half of the urine, while, if no more secretion be present in the urethra, the urine passed later must appear always clear and transparent.

As already noted, this form of pyuria is found with

the various forms of urethritis. The milder kind of urethritis has its type in urethritis catarrhalis, the severer
kind in urethritis gonorrhœica. It is not the aim of this
sketch of pyuria to discuss gonorrhœa in detail. We
will here give only so much as appears absolutely necessary to understand and treat chronic gonorrbœa (without
the aid of the endoscope).

Urethritis catarrhalis is a disease of the urethra,
which as a rule runs its course in the most superficial
layers of the mucous membrane of that part. The
exciting causes are sometimes traumatic and chemical influences, often dyscrasiæ, and diseases affect-
ing the system generally. Among traumatic causes we
may mention forced instrumental interference (catheterizing) and masturbation ; among chemical causes, the
inappropriate use of medicinal agents in the urethra, and
sexual intercourse with unclean persons, who are not,
however, affected with gonorrhœa.

Concerning the relation of urethritis catarrhalis to a
dyscrasia, we must refer to those discharges that sometimes accompany tuberculosis and syphilis, and which
resist all our efforts to cure, until we take the general
systemic disease sufficiently into consideration. Often
enough, gonorrhœal infection is combined with such a
condition, and then the course of this form of urethritis
is a very chronic and therapeutically, a most obstinate
one, which only improves by simultaneously treating the
general disease, and thus it sometimes finally gets well.
The urethritis after trauma in catheterizing, after inappropriate intra-urethral local treatment, and after im-
pure non-gonorrhœal sexual intercourse, is usually but of
short duration ; it often disappears after a few days if the
hurtful influences be removed. The course of urethritis
is not so rapid if it has arisen after masturbation has

been practiced continuously for years. In this case the urethritis yields only to an energetic local and instrumental procedure, both being carried on at the same time. The secretion of catarrhal urethritis is, as a rule, not purely purulent. It is, on the contrary, whitish or gray, and stains the linen with white spots which have rather a dark border and a central yellow point. Microscopically we find, it is true, many pus-cells, but at the same time the epithelium from the urethra is always found in large amounts. We not infrequently find such a urethritis in boys of ten or twelve years, if these are addicted to the vice of masturbation. It is very different with gonorrhœal urethritis. *Gonorrhœal urethritis* always constitutes, as compared with catarrhal urethritis, a more intense disease. In the first place, its normal course is a much longer one, and in acute cases seldom ceases before the fourth to the sixth week. Gonorrhœal inflammation of the urethra only occasionally attacks the most superficial layers of the urethral mucous membrane alone. The inflammatory irritation is more apt to extend deeper in certain places, and excites the submucous layers of the urethra to further inflammatory hyperplasia with a chronic course.

Thus, we not infrequently find, here and there, portions of the urethra affected with gonorrhœa infiltrated in its entire thickness, so that such an infiltration can be even felt by the finger. Not only the walls of the urethra are attacked by this inflammatory process, but in certain cases we find much peri-urethral infiltration, not only perceptible to the touch, but even visible as a swelling to the eye, and which often ends (if it becomes an abscess) by perforating the skin of the penis and evacuating its purulent contents. From this description it is clear that the gonorrhœal disease of the urethra is capable of getting into the deeper layers of the urethral walls, and that, in a chronic case, it may cause the most diverse alterations of the urethra itself. The superficial changes caused by the gonorrhœal process have been made accessible to the eye through the endoscope ; the deeper alterations, on the contrary, express themselves by metamorphoses of the urethral walls, sometimes taking up their entire thickness. The coats of the urethra become rigid, and since their elasticity is lost, they are together changed into a stiffer-walled, more rigid urethra, which has a slightly lessened calibre.

Otis first called our attention to these alterations in the urethra after gonorrhœa, and named them "strictures of wide calibre." A microscopic examination of urethras from individuals who have got well after a chronic gonorrhœa, generally shows two important and striking alterations. One concerns the epithelium, the other the submucous and urethral connective tissue. In isolated places the epithelium may often be seen in massive layers one over another, and show us what is called a heaping-up of the epithelium of the urethral mucous membrane; the submucous connective tissue appears also much thicker, there is more of it, and in some places it forms layers which occupy the entire thickness of the urethra.

This revelation of the microscope shows clearly enough, that as compared with the catarrhal urethritis, the gonorrhœal process is very apt to attack the deeper layers of the urethra, and there cause those alterations— the well-known consequences of gonorrhœa—strictures. Acute gonorrhœa begins at the extremity of the urethra, where the infection always takes place, and gradually extends backward from this place. In its usual course it stops at the compressor urethræ in the fourth or sixth week, and in proportion as the disease concentrates itself in the back of the urethra, the anterior, first inflamed part gradually loses its very red look. If the gonorrhœa passes the boundary of the compressor urethræ, an abnormal course of the disease sets in, not unfrequently accompanied by prostatitis, cystitis, epididymitis, etc., and thus prolongs the duration of the gonorrhœa very considerably. Now, although the gonorrhœa gradually attacks the entire urethral mucous membrane, still there are particular favorite places where it is apt to linger in the chronic disease. These places are the physiological dilatations of the urethra, the fossa navicularis and the bulbar portion. Thus in both these places we find the consequences of chronic gonorrhœa (that is to say, strictures) most often and most strongly pronounced.

To sum up, purulent discharges from the anterior urethra (as far as the compressor urethræ) are characterized by the fact that, if the urine is passed into two glasses, only the first half will always be dimmed by flakes, fibres, or otherwise, while the second part of the urine remains clear and transparent ; and further by the fact that, in the intervals between acts of urination, the secretion escapes spontaneously from the meatus, or at least appears at this place, spotting the linen to a greater or less degree, since there is no muscle in the urethra between the compressor urethræ and the meatus to cut off the free exit of pus.

II.

FROM a practical stand-point it seems best to consider
the urethra as divided into an anterior and a posterior
portion. The anterior part consists of the whole length
of the urethra as far as the compressor urethræ muscle.
The signs and symptoms of suppuration of this part
have been discussed in what has just preceded. The
posterior portion of the urethra, on the other hand, comprehends the membranous and prostatic parts as far as
the internal sphincter muscle, the so-called neck of the
bladder. The term "neck of the bladder" is denominated
unscientific by many writers as a synonym for the prostatic portion, and not justified by the anatomy of the
part. In point of fact, if we take a bladder which has
been dried and blown up, we do not find a dilatation
similar to the neck of a bottle in the prostatic portion,
opening directly into the bladder. On the contrary,
the sphincter internus is seen to be closed, and in this
way the prostatic urethra shut off from the bladder by
this muscle. In spite of this, the fact that this portion
really belongs to the bladder, or, conversely, that the
trigonum Lieutaudi belongs to the prostatic urethra, is
evident from the fact that both the lateral limbs of the
trigonum can be followed anatomically far into the prostate, and, moreover, the muscular layer of the prostate

represents simply a prolongation of the muscular coat of the bladder.

The term "neck of the bladder" has, however, a practical significance. The neck of the bladder, or the posterior part of the urethra, comprehends the portion between the internal sphincter and the external cut-off muscle, the compressor urethræ. Now the strength and resisting power of these two muscles are very unequal. While the external cut-off muscle forms a barrier to fluid pressing either toward the bladder from without, or from within the bladder outward, which is only removed by the volition of the individual, we find that the inner sphincter yields to very slight pressure, and, therefore, offers but slight resistance either to the urine in the bladder pressing outward, or to secretion arising in the posterior urethra pressing its way backward. Thus the desire to urinate, as a rule, is caused by the pressure of the urine overcoming the resistance of the internal sphincter, and getting into the "neck of the bladder."

In the moment of the strongest desire to urinate, the neck of the bladder and the bladder itself form one common cavity, and the further escape of urine is only hindered by the cut-off muscle (compressor urethræ) dependent on the action of the will.

Any one can convince himself of the unequal power of these two muscles by injecting fluids into the bladder. For example, if we try to inject the bladder simply through the anterior urethra, without using a catheter, by means of an irrigator or syringe, we find the greatest resistance from the compressor urethræ muscle, and in isolated cases, in spite of the strongest pressure, and painful distention of the anterior urethra, we do not succeed in causing the fluid to pass into the

2

bladder. This difference between the two sphincters of the urethra is seen, however, most clearly in case of disease of the posterior part of the urethra itself.

For example, if catarrhal secretion, blood, or any other fluid accumulate in this part, it will (since it is hindered by the compressor urethræ muscle) never appear in the anterior urethra as a visible discharge, or as stains on the linen, but, on the contrary, it will overcome the weaker sphincter, and enter the bladder. Thus we find, in bleeding from the prostate, the blood flows back into the bladder, tinging the urine, as a whole, as if it came from the bladder itself. Likewise the purulent secretion, in cases of prostatitis, we not infrequently see flow back into the bladder, causing it to take on inflammation. It is, moreover, a very well known fact that, as soon as a gonorrhœa in its abnormal course has passed the boundary of the compressor urethræ, the most violent symptoms connected with the function of the bladder, such as constant desire to urinate, and tenesmus, usually ensue.

It is clear enough, from the facts just mentioned, that the posterior portion of the urethra belongs more to the bladder than to the urethra, and that, therefore, the name "neck of the bladder" is justified from the practical point of view also. Therefore, in diseases of the "neck of the bladder," the urine in the bladder will be made turbid or not, according to the quantity of the secretion of the part. If only a little secretion has collected in the posterior urethra, the urine in the bladder remains uninfluenced, and if we have the patient urinate successively into two glasses, only the first portion of the urine passed will appear turbid, while the second half remains clear and transparent.

If, however, the secretion in the posterior urethra is

considerable in amount, it will flow back into the bladder, make the urine more or less turbid, and even irritate the bladder itself. In this case, both specimens of urine (passed into two glasses) will appear turbid. However, as a distinction from a primary cystitis, the first half of the urine will appear *more* turbid than the second, and will contain more compact flakes, which all come from the urethra, and which accordingly are absent in the second portion of urine passed. An additional characteristic of diseases of the neck of the bladder, and as a distinction from urethritis of the anterior urethra, is the fact that there is never any discharge from the meatus when the disease is limited to the "neck" of

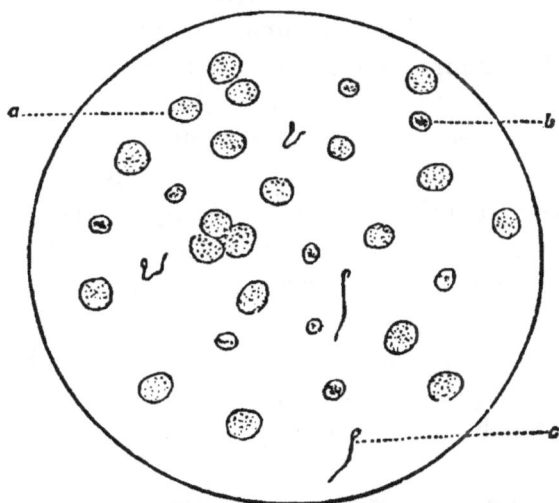

FIG. 3.—Urinary sediment from a case of catarrh of the neck of the bladder. *a.* Pus-corpuscle. *b.* Blood-corpuscle. *c.* Spermatozoön.

the bladder. The secretion of the "neck of the bladder" is sometimes copious, sometimes very scanty, and may consist of increased mucous secretion, or again of pus. If the secretion is scanty, it usually appears in the form of flakes or particles of membrane ; if more copious, it will make all the urine turbid.

Not seldom the catarrhal secretion is mingled with spermatozoa, or even blood. The admixture of blood is by no means a rarity when there is at the same time strong and painful tenesmus. In these cases, at the end of urination, the inflamed, swollen, and vascular mucous membrane is usually squeezed by the spasmodic contraction of the sphincter. Hence, we here see drops of blood only at the close of urination, which spot the linen in a characteristic manner.

The ætiological factor for the catarrh of the neck of the bladder varies widely. Gonorrhœa is the most frequent cause. Still we very often find mild catarrhs of the vesical neck after masturbation and sexual excesses. While the gonorrhœal catarrhs generally cause a purulent secretion, those coming after masturbation and sexual excesses are usually accompanied by numerous flakes and an increased mucous secretion.

No small contingent is afforded by tuberculosis of the genito-urinary tract, which may appear in the most varied forms, and is sometimes extremely difficult of diagnosis.

Finally, all primarily acute and chronic diseases of the prostate gland cause inflammatory processes in the neck of the bladder. New growths and calculi are also factors.

The alterations which are present in catarrhs of the vesical neck, and which can be determined by the endoscope, are swelling, folding, and reddening of the mucous membrane. A slight touch causes the latter to bleed readily. Not seldom we find the blood-vessels much dilated, and the mucous membrane partially robbed of its epithelial coat, so that on endoscopic examination the surface appears not unlike that of a granulating wound.

Sometimes alterations can also be detected on the caput gallinaginis, and it appears in certain cases much enlarged, hypertrophic.

The neck of the bladder is that part of the urinary tract whence tenesmus most frequently starts. So it is clear that in diseases of this part increased frequency of urination will be a constant symptom. If the catarrh be acute, the desire to urinate will be strong, and very painful at the same time. On the contrary, if the process is a chronic one, the pain is reduced to a vague and disagreeable sensation at the end of urination, the frequency, however, remaining. Frequent micturition in disease of the posterior urethra is such a very characteristic symptom that from the presence of this sign alone we can always conclude with certainty upon a lesion in the neck of the bladder.

If we investigate the different portions of the urinary tract from a pathological stand-point in relation to the desire to urinate, we find, as a matter of fact, the "neck of the bladder" is affected most often, and in a more intense degree than the other portions. Thus diseases of the kidney are only accompanied by frequent micturition, and then but temporarily, when, at the same time, a catarrhal process invades those parts of the tract which carry off the urine, as sometimes occurs in acute cases. On the contrary, in chronic nephritis, in tumors of the kidney, or in renal hæmaturia, we find this symptom wanting. Also *primary* pyelitis, which is partly a renal affection, is never associated with this abnormal frequency of micturition in distinction from an ascending pyelitis from gonorrhœa, which *has* this symptom, since the "neck of the bladder," as well as the bladder itself, is affected at the same time. Again, in calculous pyelitis, we only have vesical tenesmus when the concretions are passing off. Even very large stones may be present in the pelvis of the kidney without causing tenesmus, unless at the same time the neck of the bladder be irritated. In hydro- and pyo-nephrosis this symptom is also wanting. Even in diseases of the bladder itself, when primary in origin, and especially when the causes are to be sought in the walls of the bladder, or when the disease is situated at the fundus, there is little tenesmus present; but when the neck of the bladder becomes involved, strong tenesmus sets in at once.

This phenomenon is illustrated best by stone in the bladder. If the stone be near the fundus, as is usually the case when the patient is in bed, frequent micturition disappears, or at least is very much less; but as soon as the stone gets near the neck of the bladder, as is generally the case in the vertical position and in walking, annoying and painful desire to urinate sets in at once. In the same way, inflammation of the anterior urethra has no frequent micturition accompanying it, but it immediately follows an extension of the inflammation to the membranous or prostatic portions. On the other hand, the most annoying and painful tenesmus comes with the different forms of prostatitis, with new growths, and in tuberculosis of the prostate. The tenesmus may in these cases increase to such an extent that in acute cases the patients hold the chamber-pot in their hands almost day and night, or, in chronic cases, are obliged to wear a rubber urinal. In addition, the tenesmus is so painful at the end of urination that the patients cry out with pain. From this description it is clear that diseases of the vesical neck are usually accompa-

nied by an annoying, and in many cases painful, frequency of urination. A further symptom of disease
of the neck of the bladder is, that at the end of urination (especially if frequent), when painful tenesmus
occurs, at the same time a few drops of thick pus or
blood, or a mixture of both, escapes from the meatus.

In eminently chronic cases, semen, or a "white sand,"
is passed at the end of micturition. This "white sand"
consists principally of granular carbonate of lime, which
appears in the aggregate as spherules the size of poppy-
seeds. If we examine these spherules with the microscope, we find among them many pus-corpuscles which
seem coated with the calcic carbonate. Reflex neuroses
are a frequent phenomenon in diseases of the prostate
and neck of the bladder. These may be of purely local
occurrence, and affect the genito-urinary apparatus alone,
or they may consist of general phenomena, which show
that the entire nervous system is involved.

The local reflex neuroses show themselves sometimes
as disturbances of a sensory or motor nature, or again of
a secretory nature, which may all appertain to either the
urinary or to the genital apparatus. The reflex neuroses
of the general nervous system appear either as a strongly
increased general reflex excitability, as nervousness, or
as a greatly lessened nervous activity, as apathy or
melancholia. (See "Neuroses of the Male Genito-urinary System," Ultzmann, "Wiener Klinik," 1879).

This circumstance finds its explanation on the one
hand in the fact that the prostate (neck of the bladder)
is that part of the urinary tract richest in nerves and in
ganglia, and on the other in that the hypogastric plexus
of the sympathetic (which by means of its vesical
plexus in man supplies the seminal vesicles and the
prostate as well as the bladder) stands in direct commu-

nication by nerve-fibrils with the sacral ganglia as well
as with the pudendal plexus of the sacral nerves.

Catarrhs of the neck of the bladder are sometimes
accompanied by alterations of this part of the urinary
tract, which may be determined by digital examination
per rectum, or by investigating the urethra by means of
a sound. Thus we often find, in cases of long-standing
catarrh of the vesical neck, the lobes of the prostate
irregularly formed, flattened, or grooved; or its surface uneven, hard, and rough. Again, sometimes we find
hard, circumscribed infiltrations in one or more lobes
(chronic prostatitis).

In catarrhs of the neck of the bladder consequent
upon masturbation or sexual excesses, the entire prostate is not seldom atrophic, and, indeed, sometimes
to such an extent that we can no longer detect the contour of the separate lobes with the finger. In catarrh
of the neck of the bladder, in consequence of tuberculosis, we sometimes find the prostate irregularly uneven
and hard ; at the same time, one or the other of the seminal vesicles changed into a hard cord. Sometimes, too,
the epididymes are felt as large, hard, insensitive bodies.
But there are catarrhs of the neck of the bladder where
no considerable alteration can be detected by the finger
per rectum. Indeed, this is the most frequent result,
especially when the catarrh is not a very old one. In
examining with the sound, we find the anterior urethra
as far as the compressor urethræ normal; the walls
soft, elastic, and but little sensitive ; but as soon as we
arrive at the region of the membranous urethra and
prostatic, we feel an increased resistance, the sound
passes with difficulty, and only with an increase of pressure, into the bladder. At the same time we notice a
greatly increased sensitiveness of this part. These are

changes which find an easy and complete explanation in the chronic inflammation of the part.

We also find in these cases, and especially in the very slight catarrhs, such as usually follow masturbation, not infrequently a spasm of the compressor urethræ, which has been denominated "spasmodic stricture" by some authors. The spasm in the compressor urethræ finds its analogue in the spasmodic contractions of the circular muscular fibres of the rectum, when catarrhal or ulcerative processes are present. Not seldom we find spasmodic contractions in both rectum and urethra at the same time, where the lesion can be detected in but one of these. Thus we find spasm of the rectal muscles in cases of catarrh of the neck of the bladder, and *vice versa*. This is explained by the fact that both regions are supplied by the same nerves, viz., the middle and inferior hæmorrhoidal. Since the spasmodic contraction of the compressor urethræ which follows the introduction of a sound is most surely overcome by the use of thick, metallic sounds, it is clear that in these cases we should never use conical instruments, or those of small calibre, and always perform the introduction *lege artis* with great care. A constant pressure with a smoothly rounded and thick sound will overcome the spasm of the compressor. On the contrary, an unsteady catheterization, especially if performed with a small instrument, will only excite the muscle to increased activity. One can thus only do injury, and still not get into the bladder. The subjective sensations of the patient in catarrh of the neck of the bladder are concentrated in the perinæum, in distinction from actual cystitis, where the tenderness and pain are felt mostly above the pubes. Patients complain sometimes "as if something in the rectum were drawing itself together " ; sometimes of a sensation of fullness in the perinæum ; again, of a burning in the course of . the urethra, or of a lancinating pain in the glans penis, while pressure with the hand over the fundus of the bladder is well borne by the patient without any pain whatever.

If we did not carefully examine the genito-urinary apparatus, we might frequently make a wrong diagnosis of pyelitis from an examination of the urine, when the case was actually one of disease of the neck of the bladder. That is, such a urine has usually an acid reaction ; it contains pus, and sometimes also a larger amount of albumen than corresponds to the quantity of pus present. A microscopic examination of the sediment shows an absence of triple phosphate crystals ; and the small, round, swollen (from inflammation) epithelial cells from the neck of the bladder are not always easy to distinguish from altered renal epithelium. Since, now, in disease of the neck of the bladder, we frequently get radiating neuralgic pains in the sacral region, the deception is the more complete. The increased separation of albumen in disease of the neck of the bladder is explained by the hindrance to the outflow of urine from the ureters. This hindrance may be caused either by inflammatory swelling of the prostate and its surroundings (prostatitis and peri-prostatitis), or by pericystitis, and in women by parametritic exudations, by which a mechanical compression is exerted in the neighborhood of the openings of the ureters into the bladder. Then it may also be present when no such infiltration in this region can be detected, when severe and painful tenesmus is present, whereby a hindrance to the flow of the urine from the ureters also results. In these cases the albuminuria present is to be explained by the partial setting back of the urine in the ureters, toward the kidneys—analogous to those cases of chronic retention with insufficiency of the bladder, or where there is some other obstruction to the evacuation of the urine, as hypertrophy of the prostate, or a narrow stricture of the urethra.

This albuminuria occurring in diseases of the neck of the bladder, and especially in acute inflammatory processes here localized, is often very characteristic and pronounced in a gonorrhœa in its abnormal course.   As long as the clap is limited to the urethra in front of the compressor urethræ, the abnormal frequency of urination (Harndrang) is absent, as well as renal albuminuria. But if the gonorrhœa passes the limit of the compressor urethræ, both albuminuria and frequent micturition frequently set in at once.   The albuminuria is in direct proportion to the tenesmus.   Now, if in such cases we give narcotics, and in this way cause the constant desire to urinate to vanish, the albumen also disappears at once from the urine, or, at least, it considerably diminishes. At the same time, we find that during the tenesmus but a small amount of urine is passed, in spite of the frequent micturition, while the urine begins to flow more abundantly after the disappearance of this symptom.   This is certainly a proof that there is a sympathetic albuminuria in diseases of the neck of the bladder, accompanied by strong tenesmus, that has nothing in common with pyelitis, but which from the microscopic and chemical examination of the urine has been confounded with it.   Only when pyuria and albuminuria are present at the same time with polyuria, and when no painful micturition exists, can we infer in doubtful cases that a pyelitis is present also, or that pyelitis is the only disease at the time.

Summary.—Thus catarrh of the neck of the bladder is characterized by the fact that—first, tenesmus and sensitiveness are felt at the close of urination ; secondly, a discharge from the urethra is wanting ; and, thirdly, in the urine passed into two glasses, only the first part appears turbid, or, if both are somewhat turbid, the first is the more so of the two.

III.

SUPPURATION OF THE BLADDER.—CYSTITIS.—CATARRH OF THE BLADDER.

By catarrh of the bladder, in general, we understand a catarrhal inflammation of the mucous membrane of the bladder. This inflammation has usually the property of liberating a ferment at the same time as the catarrhal secretion is formed, which changes the urea at once into ammonium carbonate, and which immediately produces alkaline fermentation of the urine. Cystitis and ammoniacal fermentation are so well known as inseparable terms, that formerly the differential diagnosis between cystitis and pyelitis was always made by litmus-paper. To-day we know that there are vesical catarrhs with a urine of acid reaction ; and that, on the other hand, we not seldom see an alkaline urine which has no relation whatever to a catarrh of the bladder.

We may divide catarrhs of the bladder into *acute* and *chronic,* further into *partial* and *total,* according as only a part of the bladder (in the neighborhood of the opening of the urethra) or the entire bladder is involved. Partial vesical catarrh is usually a process propagated from the urethra, and one of the accompaniments of every severe catarrh of the neck of the bladder. A total catarrh of the bladder, on the contrary, is usually caused by changes which involve the entire wall of the bladder (excentric hypertrophy of the bladder—paralysis, etc.).

According to the extension in depth of the inflammation of the bladder, we may speak of a *catarrhal* or a *parenchymatous* cystitis ; in the first case, only the mucous membrane of the bladder is involved ; in the second, its muscular coat also ; or, according to the character of the inflammatory product, we speak of it as a *mucous*, a *purulent*, or an *ichorous* cystitis, according as the urine contains mucus, pus, or a grayish-red, shreddy, putrid matter. Primary catarrhal inflammation of the bladder is one of the greatest rarities. Thanks to its protected position and its difficult communication with the outer world, the bladder is quite shielded from the common causes of catarrhs of other mucous membranes. While injurious substances easily make their way into the lungs or alimentary canal, this is not the case with the bladder. When no urine is flowing through the urethra, it is a long, closed canal which perfectly hinders the entrance of anything harmful into the bladder. Then an intact vesical mucous membrane is scarcely capable of any absorption whatever. If we inject a solution (one half to one per cent) of iodide of potassium into the healthy bladder, we can soon be convinced that no iodine appears in the saliva, even after the lapse of an hour. Again, as is well known in cases of retention of urine, the bladder may be distended to a tumor the size of a man's head, and remain so for days, without any absorption of the constituents of the urine ensuing.

It is only when the bladder in its totality is involved, that is, when all its coats are more or less altered by inflammation (parenchymatous cystitis), that foreign matter from the immediate neighborhood can get into the bladder. As a matter of fact, in parenchymatous changes which involve the entire thickness of the bladder, we get a feculent urine (i. e., one which smells like fæces). If we were to mention percutaneous, catarrh-causing factors, the influence of so-called chilling, or "catching cold," would come first. This is an etiological moment brought about by unknown influences, atmospheric, telluric, or due to changes of temperature, to which we must often appeal when the real cause is not easy to discover. If we *are* entitled to assume taking cold as a cause of inflammation in those viscera most exposed to atmospheric influences, we must be very cautious in adopting this as a factor in bladder-catarrhs. By virtue of its perfect occlusion from atmospheric air, the bladder is far more removed from injurious influences than is the gastro-intestinal tract, or the respiratory organs ; indeed, it is well known that catarrh of the bladder is of extreme rarity in men who have never had any disease of the generative organs, especially gonorrhœa. Again, it seems strange that childhood and boyhood show an especial immunity from vesical catarrh, although the so-called "colds," the catarrhs of the gastro-intestinal tract, and of the organs of respiration, are very common in these periods of life. It becomes common only in youth, and then generally either as an immediate consequence of gonorrhœa, or perhaps only first appearing several or even many years after. These latter cases are the very ones which people are apt to refer to "catching cold," drinking imperfectly fermented beer, sexual excess, etc. When we examine these cases more closely, we find that at some previous time, it may be many years before, the patient had a gonorrhœa, lasting a good while, and perhaps accompanied by epididymitis or cystitis. We also find that the urine commonly contains whitish threads, that it is perhaps rendered more turbid at times by increased secretion of mucus, and that now and then, especially at the end of urination, the patient has a peculiar, uncomfortable feeling which may even amount to slight tenesmus. All these patients consider themselves otherwise as in good health, and never think this a consequence of the previous gonorrhœa, especially since that occurred years before, and on that account they have supposed themselves well long since.

But this is not the case. Gonorrhœa, and especially those forms which have penetrated as far as the prostatic portion—the deepest part of the urethra—seldom forsake this region without leaving behind traces of their presence. Such a liability to a recurrence of the inflammatory process still remains, that a sexual excess, or the use of poor beer, etc., is quite enough to set up inflammatory phenomena or catarrh at the vesical neck. If these processes extend to the bladder, a cystitis arises. Another etiological factor — although a far rarer one—is to be sought for in those low constitutional conditions closely allied to scrofula or tubercle. In such individuals these bladder-catarrhs usually get worse in the severer seasons of the year, and not infrequently vanish in summer, when the entire organism shows that it is improving under a strengthening and appropriate diet. Swellings of the epididymis—painless and chronic—not infrequently accompany this form of catarrh of the bladder.

We not infrequently find rectal fistulæ, scrofulous glandular swellings, and diseases of the bones at the same time. We can not always determine tuberculous disease of the lungs in such cases ; likewise the examination of the prostate and seminal vesicles for hard infiltration does not always give a positive result. The appearance of the individual, as if he were by inheritance anatomically predisposed to phthisis, is often our only ground for believing this the etiological factor. Moreover, daily clinical experience teaches that various acute febrile diseases, especially such as are apt to localize on the skin and mucous membranes (exanthemata), may give rise to catarrhs of the bladder. It is also well known that in consequence of lithiasis, of new growths in the bladder or in its neighborhood, and of injurious chemical and mechanical action, catarrh of the bladder may be excited and kept up.

Advanced age affords no small proportion of cases of cystitis.

The most frequent factors in catarrh of the bladder, after the age of sixty years, are to be sought for in senile changes in the bladder and prostate. Sometimes it is a hypertrophy of the prostate, sometimes hypertrophy and dilatation of the bladder, or again paralysis of the latter, which causes vesical catarrh by reason of the retention of urine; or, in fact, any obstacle to the free escape of urine may cause it. Finally, it is not to be denied that certain resins, balsams, and ethereal oils, which have an especial selective action on the genito-urinary tract, may sometimes cause catarrhs of the bladder. In view of these etiological factors which we have just considered, we are obliged to believe that a primary vesical catarrh as such, does not exist ; or at least, that it is one of the greatest rarities. It is also clear that a mere diagnosis of catarrh of the bladder is quite an insufficient one, and that we must in every case, after making such a diagnosis, look for some other cause than the very convenient one of " catching cold," or using too fresh beer or wine ; and, finally, we see, from the preceding, that it is only by exactly weighing and considering all these factors that we can begin any proper and successful treatment, whether local or general. The *diagnosis* of catarrh of the bladder is based moreover upon two symptoms usually present, in addition to other characteristic signs, and which may be very deceptive, although their great importance can not be denied. If the practicing physician does not carefully examine the urine in these cases, he may be led astray by the presence of these symptoms, and make a diagnosis of catarrh of the bladder (or cystitis) where no such thing exists. These two symptoms, most commonly misinterpreted, are "*frequent micturition*" and "*alkaline reaction of the urine.*"

Frequent micturition is no attribute of cystitis alone ; on the contrary, there are many general diseases as well as various local irritable conditions about the "neck of the bladder" which cause a very troublesome desire to micturate when there is not a trace of bladder-catarrh present. Among general diseases, there are disturbances of nutrition, which run their course with polyuria, diabetes mellitus, diabetes insipidus, hydruria, or spasm of the detrusor vesicæ in consequence of a central lesion of the nervous system. In local conditions of irritation (and it is precisely these states that are most often mistaken for cystitis) there is an especial sensitiveness of the neck of the bladder, such as not infrequently follows masturbation, sexual excesses, or occurs after a gonorrhœa has run its course, and from this a frequently occurring desire to urinate is originated as a reflex impulse. In all these cases, by a careful examination of the urine alone, we can make a negative diagnosis that no cystitis is present, because the urine in these cases is usually very clear and transparent. If there is no catarrhal secretion in a given urine, we can not very well say a bladder-catarrh is present.

The alkaline reaction of the urine is still more deceptive than frequent micturition. We must always distinguish two kinds of alkaline urine. The urine may be alkaline from the presence of fixed alkalies, or at another time it may owe its alkalinity to ammoniacal fermentation. Alkalescence through fixed alkalies, such as we usually see in phosphaturia, is easily recognized by the fact that the lime and magnesia salts in the sediment show no combinations with ammonia. In such a urine we find calcic carbonate, crystalline calcic phosphate, and sometimes also the crystalline magnesium phosphate, but we never find the great colorless crystals of ammonia - magnesian phosphate. We not infrequently find these urines in central and peripheral diseases of the nervous system, and therefore often in paralysis of the bladder. In these cases the fixed alkali is separated from the blood by the kidneys, and the bladder receives a urine, already alkaline, to store up. In quite the same way, by an unusual diet (vegetable diet), or by the plentiful use of mineral waters, or of the so-called " digestive powder " (which consists mostly of fixed alkalies), we may thus artificially get an alkaline urine with a plentiful phosphatic sediment, which, of course, has no connection whatever with a catarrh of the bladder. In all these cases it is easy, as a rule, to gradually make the urine acid by prescribing a suitable diet, and by such medicines internally as phosphoric acid, hydrochloric acid, nitric acid, salicylic acid, and especially by giving carbonic acid, in the form of pure carbonated water (a "siphon").

The carbonic acid acts in a striking way, since it very soon appears in the urine, and clears it up by transforming the lime and magnesia salts of the sediment into the easily soluble bicarbonates.

It is somewhat different when the alkalinity seems to be due to carbonate of ammonia. Here we find in

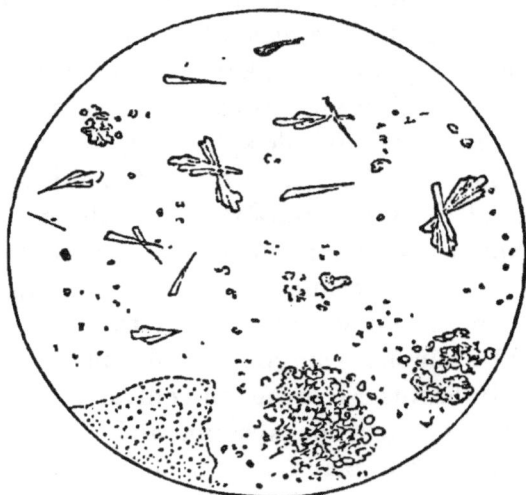

Fig. 4.—Sediment of a phosphaturia. Finely granular calcic carbonate and crystalline calcic phosphate. (Three hundred diameters.)

the sediment, along with the amorphous calcic phosphate, the crystalline ammonia-magnesian phosphate, and sometimes ammonium urate in addition ; that is to say, ammonium combinations, which make it clear that the alkalinity of the urine in question is due to ammonium carbonate.

Although we can exclude the supposition that the alkalinity may be due to fixed alkalies, as soon as we find these combinations containing ammonium in the sediment, yet this alone does not by any means entitle us to assume that a catarrh of the bladder is *necessarily* present in every such case. For example, we sometimes get a turbid and alkaline urine in the course of acute inflammatory and febrile diseases, both after the febrile stage and during convalescence, which owes its alkalinity to ammonium carbonate. The old physicians used to call this "broken urine," and explained it by supposing that the diseased matters were cast out of the body with the stinking, turbid, and slimy urine. They greeted this as a visible sign of beginning recovery. It is thus clear from the preceding statements that an alkalinity of the urine caused by ammonium carbonate can not always justify the diagnosis of catarrh of the bladder (or cystitis). In addition to this, the presence of a *catarrhal secretion* must be determined in order to permit such a diagnosis ; for there can be no catarrh without a catarrhal secretion, and therefore no catarrh in the bladder without a catarrhal secretion in the urine. Triple phosphate crystals and ammonium combinations do not make a catarrhal secretion. On the other hand, it must be acknowledged that ammoniacal fermentation of the urine is a complication of bladder-catarrh in the vast majority of cases, and therefore we can not disallow its diagnostic value in a certain degree. Corresponding to the character of the secretion, catarrh of the bladder is divided into mucous, purulent, and ichorous, as we mentioned at the beginning. A mucous or catarrhal cystitis is the least severe of the catarrhs of the bladder. Such a urine contains neither albumen nor pus. It is of a normal, wine-yellow color, and is rendered slightly turbid or cloudy by mucous secretion. The reaction to litmus is weakly acid, or neutral, according as ammoniacal fermentation is present or not. The specific gravity of the urine is normal. With the microscope we find in the cloudy sediment, along with numerous leucocytes and bladder epithelium, very many bacteria. Such a urine is very often seen in paresis of the bladder, in excentric hypertrophy of the bladder, in prostatic hypertrophy, and in many like cases where frequent and regular catheterization has been performed. Purulent catarrh of the bladder is the genuine cystitis, and the best known of all.

The urine in these cases is of a wine-yellow color, and has a distinct smell of ammonia. The turbidity is uniform and intense. The specific gravity is normal. As abnormal substances we find more ammonium car-

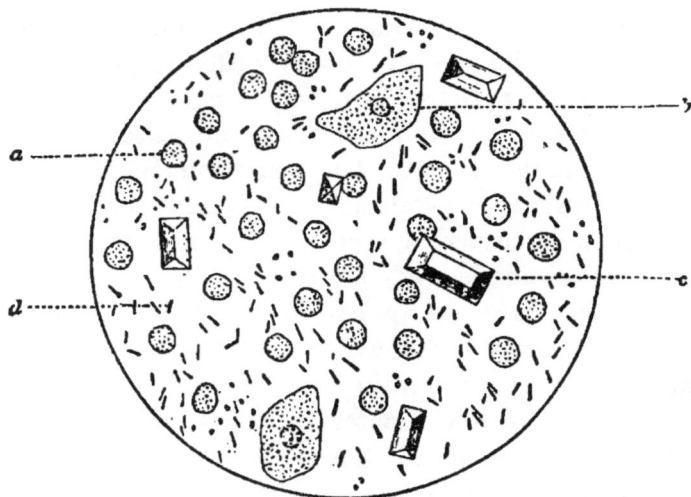

FIG. 5.—Sediment of a purulent bladder-catarrh with ammoniacal fermentation. *a.* Pus-corpuscles. *b.* Bladder epithelium. *c.* Crystals of phosphate of ammonia and magnesia. *d.* Bacteria.

bonate and albumen than corresponds to the amount of pus. The sediment is usually greenish-yellow, slimy, can be drawn out in threads, and sticks to the glass. Most of this consists of alkaline pus. Microscopically, we discover pus-corpuscles and bacteria in considerable amount; further, crystals of ammonia-magnesian phosphate and bladder epithelium. The pus-corpuscles are at the same time often greatly swollen, so that in isolated cases we can not see their contours at all.

Ichorous or gangrenous catarrh of the bladder is characterized by a brownish-green color of the urine, and by a penetrating fecal stench ; and this stench is especially developed when we mix the urine with a mineral acid. The odor reminds one of putrefying meat, of fæces, and of ammonium sulphide. The urine is very turbid, and has generally a lower specific gravity than the normal. As abnormal substances, we find ammonium carbonate in abundance, also ammonium sulphide ; further, blood coloring-matter and a great deal of albumen. The sediment is like thin pap, and will not draw out into threads. It consists of numerous bacteria, crystals of triple phosphate, of amorphous calcic phosphate, and cellular *detritus*. The cellular elements, both of blood, pus, and epithelium, can no longer be distinguished. These are all dissolved in the strongly alkaline urine, and have thereby lost their identity. Now and then only, can remnants of tissues be plainly seen with the microscope (as, for example, new growths). We find such a urine in all the diseases accompanied by extensive ulceration in the bladder, as in ulcerating new growths, in tuberculosis with formation of ulcers, in diphtheria, etc.

The ammoniacal fermentation of the urine, so often found in cases of catarrh of the bladder, takes place in the bladder itself, differing in that point from the urine of phosphaturia, which is secreted from the kidneys, already alkaline. Acid urine is secreted from the kidney, and this becomes gradually alkaline during its stay in the bladder. One may convince himself at once of the truth of the matter in a given case in this way : If we wash out a catarrhally-inflamed bladder, containing ammoniacal urine, with a solution of carbolic acid, or with such neutral saline solutions as those of sodium sulphate, sodium chloride, sodium salicylate, etc., until the washings are clear and have a neutral reaction, leave in the catheter of vulcanized India-rubber ten minutes, with the opening closed, and then test the urine which is allowed to flow from it, we find that in these few minutes the urine just secreted by the kidneys has a distinct acid reaction. This experiment shows that the ammoniacal fermentation of the urine in catarrh of the bladder takes place in the bladder itself.

Ammoniacal fermentation of the urine arises from the fact that the urea is acted upon by a ferment in such a way that it combines with water and is transformed into ammonium carbonate. Exactly what this ferment is, has not been made entirely clear up to the present time. A great many writers are devoted to the bacteria theory, and attribute the change of the urea into ammonium carbonate to the development and increase of these micro-organisms. It is not to be denied that, as a matter of fact, alkaline urine contains bacteria in considerable numbers, and especially when albumen, pus, or catarrhal secretion are present in the ammoniacal urine. At the same time, we can oppose another fact to this, viz., there are urines with a strongly acid reaction which contain both the small bacteria of one or two cells, and the long chain-bacteria. Thus, for example, the urine in diabetes mellitus never has an alkaline reaction when catarrh of the bladder is present ; on the contrary, it is strongly acid ; and yet bacteria are sometimes present in such numbers as to give the urine a peculiar glistening, prismatic appearance, like the contents of old cysts containing cholesterin. Again, the urine of those patients who are obliged to use the catheter frequently on account of hypertrophy of the prostate or paresis of the bladder is usually acid, and still it contains bacteria in considerable numbers. Since bacteria and other micro-organisms develop according to the nourishing value of the fluid containing them, it is not improbable that they are an accompaniment, and not always a cause, of the ammoniacal fermentation of the urine. This supposition appears all the more likely, since Musculus has succeeded in isolating a ferment, itself perfectly free from bacteria, which, being added to normal urine, kept at the temperature of the body, quickly changes the urea into ammonium carbonate.

As in these experiments with Musculus's ferment, bacteria soon appear after the urea is changed into ammonium carbonate, it appears very probable that, in the ammoniacal fermentation of the urine in a bladder catarrhally inflamed, the bacteria have a subordinate significance only.

This ammoniacal fermentation is, in all probability, induced by a ferment produced by the inflamed or diseased mucous membrane of the bladder itself, and which does not seem to be always of an organized nature. Probably the mucous glands of the lining membrane of the bladder produce this ferment. If the bladder is affected as a whole, patients complain of pain above the symphysis pubis, which not seldom shoots out in various directions. Pain is increased by pressure over the fundus of the bladder, and a painful desire to urinate is felt immediately. If, on the contrary, the bladder is but partially affected, and then in or near the vesical neck, no pain is felt on pressure above the symphysis. The disagreeable sensation is concentrated in the perinæum, in the rectum, and along the course of

3

the urethra, as has already been described in detail. In acute bladder-catarrh, fever is sometimes present, yet, in the majority of cases, there is no rise of temperature or increased frequency of the pulse. In chronic bladder-catarrh, the pain is considerably less as soon as the urine is passed, if stone or new growths are not the cause of the catarrh ; but the intolerance of the bladder to its proper contents, the urine, still persists, as frequent micturition shows.

The pathological appearances vary greatly according to the degree of inflammation. In acute cystitis of brief duration, all the tissues soon resume their normal condition. If the cystitis has lasted longer, the mucous membrane is usually found hypertrophied, and its veins enlarged. In acute cystitis, the mucous membrane shows only slight swelling and reddening. Sometimes this reddening is due to overfilling of the smallest vessels in the mucous membrane ; sometimes to puncti-form extravasation. These appearances are usually most pronounced around the neck of the bladder and near the trigonum. Sometimes the inflammatory process in the mucous membrane assumes a croupous or diph-theritic character. In these cases we find grayish-white false membranes in circumscribed patches (sometimes consisting of necrotic mucous membrane, again of fibrin) on a deep-red background of mucous membrane. The size of these false membranes occasionally exceeds that of the palm of the hand ; they are, how-ever, usually much smaller. On the surface which was toward the bladder-wall are seen dots of a red color, like ecchymoses. These pseudo-membranes often have a thickness of several millimetres. Moreover, in paren-chymatous cystitis, we occasionally find abscesses in the bladder-walls, which may attain considerable size. A consequence of inflammation in the muscular layer of the bladder is, that the bladder sometimes loses its elasticity and shrinks. Such a bladder—also called cicatricial bladder by the French—has but a very small capacity, and frequent micturition is a necessary consequence of this condition, even in those cases where the bladder-catarrh seems to amount to very little.

Trauma, tuberculosis, and finally new growths, may modify the condition of the bladder in a characteristic way.

Real catarrh of the bladder is distinguished from catarrh of the "neck of the bladder" in that, first, ammoniacal fermentation of the urine is generally present at the time; second, the painful sensations of the patient are localized more above the symphysis, in the body of the bladder itself; and, third, the urine being passed into two glasses, the first half appears just as turbid as the second half.

IV.

SUPPURATION IN THE PELVIS OF THE KIDNEY, AND IN THE KIDNEY ITSELF. — PYELITIS AND PYELO-NEPHRITIS.

PYELITIS, purulent catarrh of the pelvis of the kidney, is partly an affection of the kidney also, since the papillæ of the kidney which open into the calyces are usually involved at the same time. So it seems justifiable to substitute the term pyelo-nephritis for that of pyelitis. Some authors make the distinction between pyelitis and pyelo-nephritis, that in the latter casts should be present in the urinary sediment. Since, however, in the same case, casts may be present at one time and absent at another, and since casts appear mostly in the acute exacerbations of pyelitis, and, finally, since a so-called interstitial nephritis complicates the suppuration from the pelvis of the kidney—a form of nephritis in which we do not constantly find casts—it seems better to drop this distinction, and to consider a pyelitis as involving the kidney at the same time—in other words, as a pyelo-nephritis. The etiological factors in pyelitis may be various. We can, however, distinguish two forms. In the one case it is primary, the disease originates in the pelvis of the kidney itself ; the other form is propagated from another source (by continuity), is secondary, and is the form which has arisen from other pathological processes, in the urinary tract, or in its neighborhood. Primary pyelitis never has very frequent micturition as a consequence, and urination is very seldom painful or causes disagreeable sensations, while the pyelitis which is propagated from the bladder or prostate usually *is* accompanied by frequent micturition, since an affection of the neck of the bladder exists at the same time. Primary pyelitis comprehends those forms arising from retention of urine in the pelvis of the kidney. These are the cases arising from sharp bends or twists of the ureters such as are apt to occur in cases of movable kidney. We must also consider the cases where a fold of mucous membrane forms just where the pelvis of the kidney passes into the ureter, in the cases where the ureter is compressed, and all those factors leading to dilatation of the pelvis of the kidney, hydro- and pyo-nephrosis. Renal calculi cause a large proportion of the cases of pyelitis, likewise tuberculosis of the kidneys, and in rare cases we see as causes entozoa (echinococci) and new growths in the kidney. Pyelitis is also sometimes an accompaniment of the various acute febrile diseases (as typhoid fever), and it occasionally complicates chronic disturbances of nutrition (as diabetes mellitus). In all these primary forms frequent micturition is absent, and it only sets in when the process gradually extends to the bladder and its neck, and it vanishes at once when the process leaves the two regions last mentioned. This phenomenon is best illustrated by calculous pyelitis. Pyelitis calculosa shows no frequent micturition as long as the calculi remain quietly in the pelvis of the kidney, but as soon as these stones get into the bladder, whether with or without colicky pains, they irritate that viscus by their presence ; frequent micturition sets in at once,

and lasts until the small stones have been all eliminated
by the stream of urine.   When this has taken place the
frequent urination immediately ceases, in spite of the
fact that the pyelitis may be found to exist just as much
as before.   This is quite as true of other forms of primary pyelitis when the inflammatory irritation is propagated from the pelvis of the kidney to the bladder.

On the other hand, we understand by secondary pyelitis either those forms which have arisen by extension
from another part of the urinary canal somewhere be-
low the pelvis, or else such forms as occur from disease
about the kidney, or near the pelvis or bladder.   The
pyelitis arising by extension of disease from the bladder
or prostate is distinguished by the complication of frequent micturition, which does not always occur in the
pyelitis which comes by extension from inflammatory
processes in the neighborhood.   Thus we find pyelitis
especially frequent after gonorrhœa and its sequelæ,
stricture and prostatitis ; further, in hypertrophy of the
prostate and in paresis of the bladder, especially when
catheterization has unpleasant consequences.   Moreover,
we sometimes get a pyelitis with perinephritis in pericystitis and in parametritic exudations, and in all those
forms of disease which in any way compress the ureters.
Finally, it is a well-known fact that severe, purulent, or
ichorous catarrh of the bladder, such as occurs in cases
of stone, new growths, or diphtheritic processes, may extend to the pelvis of the kidney or to the kidney itself.

The pathological anatomy of pyelitis varies greatly
according to its causation.   We find the blood-vessels
of the mucous membrane dilated in the catarrhal form.
The mucous membrane has at the same time a dark-red
color.   In chronic cases it is grayish, pigmented, and
covered with a purulent catarrhal secretion.   In the

diphtheritic forms we see yellow spots of necrotic tissue on the mucous lining of the pelvis. In acute cases, the kidneys themselves look hyperæmic and slightly enlarged. On section, we frequently see a whitish striation which radiates from the cortical toward the papillary region of the kidney. In an advanced stage we find little punctiform collections of pus which run together into larger abscesses. This is the suppurative interstitial nephritis,* which is not an uncommon cause of death after operations in the bladder. In many cases of pyelitis we find the pelvis greatly dilated, and the papillary region of the kidney beginning to break down into pus. Great abscesses of the kidney form in this way, whose walls are made by the remnant of the renal parenchyma and the dilated pelvis. These changes are not uncommonly found in those cases where the pyelitis must be considered as a disease of continuity, from an obstinate suppuration in the bladder. The dilatation of the pelvis of the kidney takes place here all the quicker if it is complicated by retention of urine. In isolated cases we find the urinary tubules, especially the final, straight tubules, quite filled by bacteria and cocci. There is no doubt at all that these micro-organisms come from the bladder in most cases. At any rate, an ichorous gangrenous cystitis is usually present at the time. Klebs has named this form of nephritis—where we find cocci-emboli in the urinary tubules—pyelo-nephritis parasitica.

If calculi, entozoa, or new growths are combined with pyelitis, the pathological anatomy of the case will be correspondingly modified, or even essentially changed. The symptoms of pyelitis are not very characteristic, but the information derived from the microscopical and

---

* The so-called "surgical kidney."

chemical examination of the urine is of the highest importance, and, in most cases, is sufficient for the diagnosis. In acute cases fever is present, and frequently begin with a chill.

In the chronic forms, on the other hand, all febrile action is wanting. Pain over the region of the kidneys is only observed in acute cases, or in those chronic forms (usually unilateral) that occur with renal colic, and retention of urine or purulent secretion in the pelvis of the kidney (pyelitis with renal calculi or with movable kidney).

Chronic pyelitis, as it is usually found, is due to the extension of a process from the bladder, and but seldom shows tenderness on pressure in the region of the kidney. Pain or tenderness on pressure in the small of the back is very often mistaken for pain in the kidney. It is well known that diseases of the pelvic organs are often accompanied by pain in the region of the sacrum. Thus, in uterine disease, in rectal hæmorrhoids, and especially in diseases of the prostate and the male urethra, strongly-developed sacral pain is not at all rare. However, these pains are not situated at all where the kidney is located, but are much deeper, and extend outward from the sacrum along the iliac crest toward the right or left, or even toward both sides. The region of the kidney is at the same time quite insensitive to pressure. There are usually reflex pains which radiate from the affected pelvic viscera toward the spine.

In renal abscess, chills, continued fever, a small, quick pulse and delirium are the order of the day. These vanish at once when the pus empties itself into the urinary canal, and begin again anew when pus accumulates and can not escape. This play between exacerbation and remission may last for weeks, and result in recovery or death. If the patient can not recover, the quantity of urine diminishes gradually to anuria. The extremities become cold, the skin livid in color, the pulse quick and thread-like. The tongue is dry, covered with brown crusts, and has a bright-red edge. A distressing hiccough sets in, which increases gradually, and lasts until death. Somnolence or lethargy is present at the same time, and in isolated cases the patient seems to be in a profound sleep. As the secretion of urine diminishes, obstinate vomiting sets in of brownish-green hæmorrhagic masses from the stomach, sometimes frequent fluid stools are voided, and twitching of the muscles is not uncommon. Dropsy* is never present, and only in very chronic cases do we find a little œdema about the ankles which may extend as far as the knee. These are phenomena which find their explanation partly in the retention of urinary constituents in the blood, partly in the resorption of an ammoniacal urine loaded with septic substances from ulcerative processes in the bladder. In isolated cases, uræmic symptoms are the most prominent ; in others, those of ammonæmia ; and again, in others, symptoms of pyæmia and septicæmia predominate. Some authors call this condition urosepsis. We get our most important help in the diagnosis of pyelitis or pyelo-nephritis from the urine. While oliguria (abnormally small secretion of urine) is most apt to be present in acute pyelitis, we always get polyuria as a characteristic sign in chronic pyelitis. Polyuria, in chronic pyelitis, is explained partly by the usual presence of hypertrophy of the heart, like that found with " cirrhotic kidney," and partly because the osmotic processes, the normal relations between the blood and the water of the urine as it passes along the urinary tubules,

* Ascites.

are disturbed, the papillary and medullary regions of
the kidney being especially involved.

Thus polyuria is a constant phenomenon in chronic pyelitis, and the urine secreted may amount to six
litres or more in twenty-four hours. The average
amount in this disease is, however, three to four litres.
Acute pyelitis is usually accompanied by blood in the
urine. In chronic pyelitis it is only present when
stones or neoplasms are the cause. In the ordinary
catarrhal form of chronic pyelitis, blood-corpuscles are
never to be found in the sediment of the urine. The
specific gravity of the urine is diminished according to
the polyuria. In isolated cases of pyelitis the specific
gravity is so low that from this alone retention of the
constituents of the urine may be diagnosticated. The
urine is of light-yellow color, and is rendered milky by
the pus-cells suspended in the fluid. In *primary* pyelitis the reaction to litmus of the urine is generally acid,
but if the pyelitis be *secondary* to, and caused by, a
cystitis, then the reaction may be neutral, or even de-
cidedly alkaline. The urine of pyelitis always contains
albumen in considerable quantity, and more than the
amount of pus warrants. This is conceivable if we
recollect that the pyelitis represents in part a process in
the kidneys also. If we may say that in cystitis pyuria
is present, we are justified in saying that in pyelitis
pyuria *and* albuminuria are present ; that is, in pyelitis
there is more albumen than corresponds to the amount
of pus in the sediment. To determine this difference is
no easy task for a beginner, though for one experienced
in these matters the decision as to the value of a quantity of albumen found in the urine offers no difficulty
at all. The amount of albumen in the urine of chronic
pyelitis does not generally amount to one fourth of one

per cent. The sediment of the urine consists of pus in
fine flakes. With the microscope we sometimes see the
pus-cells aggregated to short and thick cylinders, which
come from the papillary ducts of the kidney, and are
of great diagnostic importance. Sometimes we find a
renal epithelial scale, as it were, baked into the pus-
casts ; moreover, isolated epithelial cells from the urinary tubules ; these come from the large main ducts.
In calculous pyelitis we usually find in addition blood-
corpuscles. These are in the form of little globules,
such as are very characteristic of parenchymatous or
very slow bleeding from the interior of the urinary
tract, since their shape shows the destruction of the red

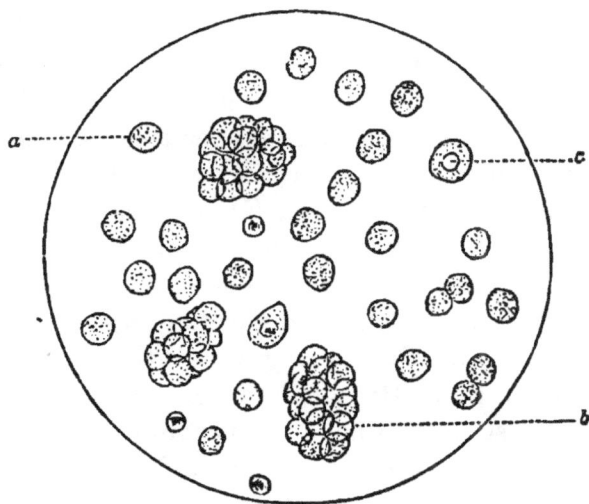

Fig. 6.—Sediment of a purulent pyelitis. *a.* Pus-corpuscles. *b.* Plugs
of pus from the capillary ducts of the kidney. *c.* Renal epithelium.
(Three hundred diameters.)

blood-corpuscles by the warm urine (urea). It is not
very uncommon to find among the cellular elements a
crystalline sediment, which may have caused the cal-

culous disease, such as uric acid, in the form of precipi-
tated crystals ; further, oxalic acid, cystin, and calcic
carbonate, or crystalline phosphate.

We also find blood-corpuscles in tuberculous pyelitis, besides molecular *débris* and bacteria in abundance. Crystalline sediments are not seen here, as a
rule. With the pyelitis caused by neoplasms or entozoa
there is usually considerable blood, but there are no
other diagnostic points to aid us, unless a piece of the
neoplasm itself, or the entozoa (echinococci) come
away with the urine.

When a chronic pyelitis has exacerbations with
fever, we find during the first few days short and thick
casts in the sediment, which are somewhat granular,
and come from the large gathering urinary tubules.
However, as soon as recovery begins, these casts vanish
entirely. If an ichorous catarrh of the bladder exist
at the same time, we sometimes see thick casts in the
sediment, consisting of bacteria and cocci from the
large straight tubes, such as Klebs describes as characteristic of pyelo-nephritis parasitica.

If the pyelitis is limited to one kidney, and occurs
when the ureter is obstructed by plugging up, or twist-
ing on its axis, we find during the entire time of such
obstruction a perfectly clear and normal urine, which
is from the sound kidney. This confirms our diagnosis of unilateral disease of the kidney. With this
retention of urine, pain is felt in the region of the pelvis of the kidney, which increases with the increase of
tension until it is unendurable, and is accompanied by
nausea and vomiting. By palpation of this region we
feel deep down, a tumor of large or small size, in which
we may even detect fluctuation at times. Examination
is, however, difficult, on account of the extreme tender-

ness. When the retention of urine in the dilated pelvis
has continued for days, the ureter may become pervious,
and, while the urine has been clear for some time,
it becomes all at once very purulent, and the tension
as well as the pain over the affected kidney vanishes.
Such retention of urine is very apt to recur ; the attacks become more frequent, the pelvis becomes more

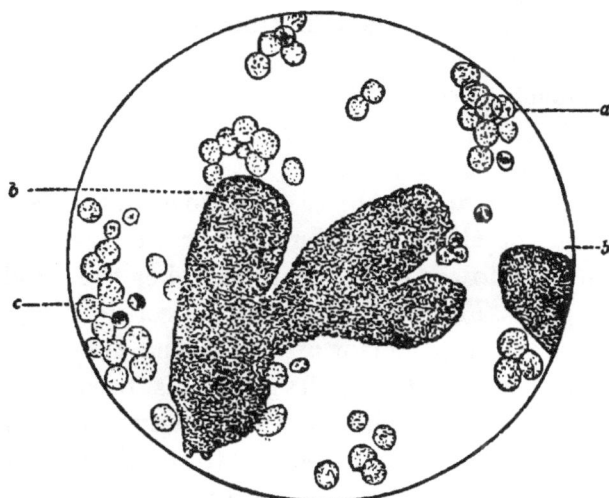

FIG. 7.—Sediment of a pyelo-nephritis parasitica. *a.* Pus-corpuscles.
b. Casts consisting of cocci. *c.* Blood-corpuscles. (Three hundred
diameters.)

and more dilated, and the ureter less and less pervious.
After some time, it may be after months, during which
time these phenomena have been repeated with more
or less intensity, suddenly a different and very painful
attack begins, which seems to last longer than the ordinary ones. The tenderness on pressure over the tumor,
as well as its size, increases greatly. Finally we may discover, by pressing with the finger, a slight œdema in the
skin of the back, over the situation of the quadratus

lumborum muscle on the affected side. The inflammation has extended from the diseased pelvis of the kidney, from the pyo-nephrosis to the connective tissue
surrounding the kidney, and here causes a peri- or paranephritis. The perinephritis now gradually causes a
purulent exudation, and a large abscess forms, which
communicates with the pyo-nephrosis. In this way concretions may escape from the pelvis of the kidney into
the perinephritic abscess, and begin to make their way
out. If the inflammatory process invades the ileo-
psoas muscle, we see symptoms of psoitis at the same
time, and the patient lies on his side in bed with the
thigh and leg much flexed.

If, in this stage of the disease, no outlet is made with
the knife for the pus—if, moreover, no surgical interference be allowed—Nature undertakes the elimination
of the pus. The abscess grows larger and larger, the
œdema over the posterior aspect of the tumor becomes
more extensive, until, finally, the pus breaks through.
The usual way is for the pus to get between the muscles
and the skin, and then to perforate the skin of the back
in the region of the quadratus-lumborum muscle. This
is the shortest and the best way. Sometimes renal calculi are eliminated in this manner, after which a spon-
taneous cure may ensue. However, the abscess may
sink down toward the abdominal cavity, or appear in the
neighborhood of Poupart's ligament, or even through
the inguinal canal. Another way for the pus of perinephritic abscesses to escape, is through the lungs. It
is relatively an unusual way, and on account of the constant danger of suffocation a more dangerous one. I
have, however, observed two cases, both of which finally
recovered after a tedious illness, and an escape of pus
through the lungs, lasting several weeks. In one case,

especially, pus was thrown off through the lungs at
the very beginning, in such quantities that it seemed
as if the patient must suffocate with each expectoration.
Finally, the pus may make its way into the intestinal
tract and may be carried off by this outlet with the
fæces. In addition, pus may empty itself into the peri-
toneal cavity ; but this is, fortunately for the patient, a
rare occurrence. The adhesive inflammation about the
abscess provides a more favorable way for the elimination of the pus. Para- or peri-nephritis is sometimes also
primary, and then usually follows an injury, but it is
much more often secondary, and due to an inflammatory process propagated from the kidney or its pelvis, as
has been above described.

The acute exanthemata and typhoid fever are occasionally though rarely the cause of paranephritis.

The urine in cases of paranephritis varies greatly
according as the process is primary or secondary to a
disease of the renal pelvis. In primary paranephritis
the urine exhibits merely the characters of "febrile
urine." It is high-colored, though perfectly clear and
transparent. There are no cellular elements in the sediment. Sometimes we may in such cases detect a small
amount of albumen. But in a paranephritis which arises
from a pyelitis or pyo-nephrosis, the urine shows all the
characters of pyelitis or pyelo-nephritis above given, unless the ureter is absolutely impermeable. When the
pus from the abscess escapes partially with the urine, we
occasionally find those dark, granular, nucleated cells in
the urinary sediment which are pathognomonic of abscess-pus.

V.

THE therapeutics of pyuria varies according as the process is an acute or a chronic one, and according to its location. Apart from acute urethritis—which is generally treated by injections at the same time that other means are employed—all experience teaches that acute pyuria and, above all, that of the bladder and neck of the bladder, ought never to be treated locally, but by internal medication and regulation of diet; while in chronic pyuria of any part of the urinary tract an appropriate local treatment takes a prominent place.

a. Therapeutics of Suppuration of the Urethra as far as the Compressor Urethræ Muscle.—Suppuration in this region includes the different varieties of urethritis. It is not the purpose of this work to give a complete review of the treatment of gonorrhœa, especially of the acute form. Neither shall we go into the endoscopic treatment. Those who wish instruction in this branch will find everything described minutely enough in the text-book of Zeissl, and further in the publications of Auspitz, Grünfeld, Gschirhakl, and others. On the other hand, we intend to give in detail the local and instrumental treatment of chronic gonorrhœa and its sequelæ, such as has done the author the best service, and such as can be easily carried out by every practicing physician.

Catarrhal Urethritis.—That form of urethral inflammation which follows trauma and injurious chemical
influences needs only suitable dietetic treatment for its
cure. Care and cleanliness of the affected urethra are
the most important things. A weak astringent solution
of alum, zinc, or potassic permanganate hastens the healing process considerably.

Gonorrhœal urethritis, under ordinary conditions,
runs its course within four to six weeks. In this time
the inflammatory process, beginning at the meatus, extends backward and stops at the compressor urethræ
if no further complications occur. It is, however, not
always entirely over in these four to six weeks, since
numerous little shreds may still be seen in the urine,
and, although the secretion has so far abated that
nothing abnormal appears to escape from the meatus,
the process plainly continues, though to a slight extent
only. And these gonorrhœal threads in the urine, after
a clap has apparently ceased, are no great rarity.

The object of treatment in acute gonorrhœa is to
secure a cyclical, natural course, so that it shall cease,
without further complication, at the proper time. It is
also to prevent (as far as possible) the inflammation
from extending further than the bulbar portion of the
urethra, into the parts surrounded by the compressor
urethræ, and involving any of the urinary or genital
tract beyond. Unfortunately, the surgeon can not always hinder this extension. Individual peculiarity of
the parts, morbid or even remote changes in the urinary
tract, have much influence here. Thus, we not seldom
see a gonorrhœa run a perfectly normal course in cav-
alrymen, who, at the same time, have to discharge their
usual exercises, while in other cases the most unpleasant
complications occur in spite of the greatest care. The

physician has, however, done his duty when, after the
medical treatment, he has carefully regulated the patient's manner of living. The first clap is usually the
most painful, as patients themselves will tell you, while
subsequent attacks no longer show such violent inflammatory phenomena. It is also a well-known fact that
successive claps are apt to follow the same course as the
first. Thus, if the first gonorrhœa was followed by cystitis or epididymitis, it is quite likely that a second or
third clap will have the same complications. So, in such
cases, it is well to give especial attention to this point.

The dietetic precautions for the patient are, that he
must keep as quiet as possible, and live very frugally.
If it is necessary to go about considerably, a suspensory
bandage is indispensable, and especially during the
third and fourth week ; that is to say, at the time when
the inflammation has reached the neighborhood of the
compressor urethræ. Regarding food, a milk diet is
very advisable ; likewise vegetables, fruit, or vegetable
diet altogether, have a beneficial influence on the course
of the disease. It is best to restrict the eating of
meat as much as possible. The use of spices, of wine
and beer, as well as of all effervescing and alcoholic
drinks, is to be avoided. If the inflammation runs high
in the first week, and if the urethra is very sensitive,
cold applications to the penis are highly beneficial.

The treatment of gonorrhœa by internal medication
has no great value. Since gonorrhœa is a local disease
of the urethra, and has no constitutional symptoms (as
is the case in syphilis), general treatment by internal
medication appears irrational. In addition, the use of
large doses of copaiba, cubebs, and other similar drugs
is by no means a matter of indifference to the organism.
It is not alone the stomach and digestion that suffer

from the use of these things, but, in isolated cases, albuminuria is also caused by such employment in large doses. Some observers think that gonitis is more apt to occur in those cases of gonorrhœa where large doses of copaiba have been used, than where a purely local treatment was employed. Small doses of these remedies have no effect, since, when they at last get into the urine, the dilution is too great to exert a healing effect upon the diseased urethral mucous membrane. Large doses, on the contrary, act injuriously in various ways. Hence it is better not to use internal medication in gonorrhœa at all, and to employ only local treatment. If copaiba, cubebs, oil of sandal-wood, and all these other remedies act so beneficially upon the diseased urethral mucous membrane, if they appear in the urine and affect the mucous membrane by coming in contact with it during urination, why do we not choose the shortest way and bring these drugs, in the form of an injection, at once into direct contact with the diseased part?

Since an acute clap always begins in front at the meatus, and only gradually extends backward toward the bulb, it is a very enticing thing to treat it locally in the first instance with great energy, and so, if possible, to nip the whole process in the bud. Favoring this idea, various abortive curative procedures were tried in former times, but these have none of them been found of sufficient value and are given up. Yet it remains for the future to institute trials of other abortive methods.

If it should appear true that the gonorrhœal infection is due to an actual transplanting of specific micro-organisms, the idea of an abortive method with anti-bacterial agents could be at least entertained. At any rate, Watson Cheyne has very recently described an abortive method consisting in the use of iodoform pessaries which shows very favorable results.

The method of treatment of acute gonorrhœa in vogue at the present time consists in regular injections into the urethra.

If the inflammation of the urethra is at the very commencement a severe and painful one, even weak astringent solutions are not well borne. In this case we inject either cold water or a one-tenth per cent solution of carbolic acid into the urethra. If, however, the urethra is not very sensitive, the use of mild astringent solutions may be begun at once. The following solutions are the best for this purpose :

℞ Alum. crud., zinci sulphat., acid. carbolic., āā 0·30=4·6 grains.
   Aquæ.............................. 200·00=6¼ oz. (about).
      M. Ft. sol.

Or else—

℞ Potassii permanganat................ 0·02=⅓ grain.
   Aquæ destil....................... 200·00=6¼ fl. oz. (about).
      M. Ft. sol.

We inject one of these solutions into the urethra three to six times a day, according to the amount of discharge. To do this properly, we have the patient urinate, just before taking the injection, in order to free the urethra from purulent secretion, and then inject at once the solution, two to four times. In the beginning inject but half a syringeful of the solution, and let it flow out immediately. Later on, in the second and third week, a whole syringeful of the solution can be injected with ease, and the meatus urethræ closed for one to two minutes by the fingers. The introduction of any instrument of the nature of a catheter during acute gonorrhœa is not advisable. By such a procedure the inflammatory process is but increased and driven backward along the urethra all the faster. According as the amount of discharge and the sensitiveness of the urethra decrease, the strength of the above injections may be increased two and three fold, and so remain until the end of the gonorrhœa. If it seems preferable, we can use tannin in the later stages, either by itself or combined with alum or some other astringent.

Again, certain mixtures holding the drug in suspension as a precipitate, or a mixture of zinc sulphate and plumbic acetate. These mixtures have the disadvantage that the fine powder sometimes stops up the ducts of the minute glands of the urethra, and in this way causes small follicular abscesses of the urethral mucous membrane. After using this kind of injection we see (sometimes even after many days) the precipitate, rolled up into little cylinders, eliminated with the discharge. Therefore it seems, on the whole, better to employ only clear filtered solutions in the treatment of acute gonorrhœa. If painful erections or chordee occur in the course of the disease, we must begin an appropriate symptomatic treatment at once.

If a clap has lasted longer than eight or ten weeks, we call it *chronic gonorrhœa*. Chronic gonorrhœa is caused (principally) by the fact that the mucous membrane, in isolated places, has not covered itself with epithelium, and for this reason these patches continue to secrete.

Patches affected in this way sometimes present the appearance of a fine granulating surface on the mucous membrane, sometimes like a superficial ulceration of the same. According to Grünfeld, small polypoid excrescences are occasionally a cause of chronic gonorrhœa.

Chronic gonorrhœa of the anterior urethra is characterized by the fact that, when the urine is passed into two
separate vessels, the secretion of the gonorrhœa—the
" gonorrhœal threads "—are always found only in the
first urine passed, while the second half appears quite
normal, i. e., clear, transparent, and free from these
"threads." Since in chronic gonorrhœa the process is
not always confined to the superficial layers of the
mucous membrane, but, on the contrary, very often in-
volves the deeper layers also, it does *not* answer to
employ simply astringents or caustics, and to try to
effect a cure in this way. For if a chronic process is
thus brought to a close, the deeper layers may still remain cicatricially changed, and from this results a
rigidity of the walls of the urethra, or a cicatricial contraction in isolated places which has unpleasant after-
effects. If we wish to obtain the most perfect cure,
we must at the same time bring the lumen of the urethra back to the normal, and the walls of the tube
itself ; this can never be brought about by simply
local application of medicaments. The normal human
urethra is a soft and elastic tube, which will easily
allow the passage of sounds of the diameter of 30
(Charrière) and larger. Such sounds are not usually
employed, because the external orifice of the urethra
seems more or less contracted. If we slit up this orifice
sufficiently, as is now often necessarily done in the operation of litholapaxy, any one can convince himself
that instruments of calibre 30, Charrière, and larger,
can be easily passed into the bladder. Otis deserves
the credit for calling especial attention to these relations and introducing large steel sounds.

If, now, in a case of chronic gonorrhœa, we slit up
the orifice of the urethra, where it will not admit of No.

30, Charrière, and then test the urethra with the large
steel sounds, we find that the affected parts do, as a
matter of fact, slightly narrow the lumen, and either
do not let the large sound pass at all, or only with some
difficulty. Otis calls these alterations of chronic gonorrhœa strictures of large diameter. Such slight narrow-
ings of the diameter of the urethra, or the loss of the
elasticity of the urethral walls, can never be discovered
by the use of small instruments, much less cured. If,
however, we wish to bring about a perfect cure, we
must not only heal the process going on at the surface,
but we must preserve the normal diameter of the urethra at the same time, and restore the elasticity of the
previously diseased urethral walls. If this last be not
done, or insufficiently done, the altered urethral walls
exert a continual irritation upon the peripheral endings
of the urethral nerves, and, as a result, we have either
those neuroses of the genito-urinary sphere so often occurring after gonorrhœa, or that slight hypersecretion
issuing from the meatus, falsely called prostatorrhœa,
and which seems to be simply a consequence of the difficult circulation of the blood through a rigid wall of
the urethra.

Looking at the situation from this stand-point, the
use of large sounds in the treatment of chronic gonorrhœa is a *sine qua non*.

In many cases this treatment of chronic gonorrhœa
by means of sounds is sufficient for a perfect and permanent cure. The "sound-treatment" is carried out
by the use of slightly conical, heavy, metallic sounds,
in the following manner : Begin with the smaller diameters, and very gradually go up the scale of Charrière,
each day or every second day using an instrument one
number larger in the series. When a sound is intro-

duced, let it remain quietly in position a few minutes. In this way we gradually rise in the scale to the numbers 27, 28, 29, and 30. Urethras which do not admit of No. 27 at least, must be carefully slit up along the frænum. If the sounds have been skillfully introduced, we very soon see the purulent secretion begin to diminish, and the gonorrhœal threads to disappear from the urine. This last circumstance is the surest guarantee of a speedy and perfect cure. If treatment with sounds alone does not suffice, we may employ a second treatment at the same time, that is, the use of various medicaments locally. After the sound has remained in the urethra a few minutes it is taken out, and the second procedure employed, so that two distinct modes of treatment are brought into use at the same sitting.

We may say here that in this case the sounds or other instruments must be smeared with glycerine, since, if the urethra-walls are covered with a layer of oil, watery solutions of astringents and caustics can not adhere to or moisten them ; and so the influence of the various medicaments upon the urethral mucous membrane is quite uncertain. We may employ local medication in a fluid, semi-fluid, or solid form. If in a fluid form, we may use them in a dilute or concentrated state. In a dilute form, we use to the best advantage the astringents as a large injection or irrigation. For this purpose we usually employ a quantity of 300 to 400 grammes, which is caused to flow slowly along the urethra.

This so-called *deep injection*, or *irrigation of the anterior urethra*, is best done in the following manner :

The patient stands ; a soft Mercier's catheter of No. 14 calibre (Charrière), with two lateral openings, is passed in as far as the bulb of the urethra. The patient holds the catheter with his left hand and a pus-basin or other vessel with his right. Then we let the medicated solution flow in slowly, either by means of an irrigator or a hand-syringe. The solution first passes out of both openings of the catheter into the bulbus urethræ, and, since it can not pass backward into the neck of the bladder (and so into the bladder itself, being prevented by the compressor urethræ muscle), it turns about, washes out the entire anterior portion of the urethra, and finally escapes from the orifice of the urethra, by the side of the catheter, into the basin held beneath.

The advantage of this irrigation is, that the medicated solution comes in contact with the bulbus urethræ —the favorite seat of chronic gonorrhœa—with some little force ; it strikes it first of all, and in sufficient quantity. This is a perfectly painless procedure.

We may employ various astringents in the irrigation. The following solutions can be recommended :

℞ Alum. crud., zinci sulphat., acid. carbolic.... āā 1·00–2·00
   Aquæ destillat................................ 400·00
     Ft. sol.

Or else—

℞ Potassii permanganat........................ 0·20–0·50
   Aquæ destillat.............. .... 400·00
     Ft. sol.

The irrigation may be done once a day.

A second way of using medicated solutions is in the concentrated form. Since these act partly as caustics, they can be employed only in a small quantity at a time, as drops. For this purpose a camel's-hair brush is best. The brush arrangement for the anterior urethra consists of three pieces, which are all made of hard rubber : *A* is a straight endoscopic tube, *B* is an obtu-

4

rator, and *C* a removable brush and handle. A drawing
of this simple apparatus makes any further description
unnecessary. The calibre of the tube is Nos. 20 to 22
(Charrière). By means of the small screw on the
shaft of the brush-handle, the brush itself may be pro-
truded half or its entire length from the tube. The
way to use it is as follows : The patient lies flat on his
back, the endoscopic tube is provided with its obturator,
smeared with glycerine, and *lege artis* introduced as far
as the bulbus. The obturator is now removed, and the
brush, impregnated with the medicated solution, is put
in its place. Now we take both brush and tube in one
hand, and by a rotary movement wipe out the bulbus.
Then we may remove the brush, dip it once more into
the solution, and again swab out the bulbus or any
other suspicious place. We may even wipe out the
entire urethra by a combined rotary and withdrawing
motion.* Any one fond of endoscopy can examine
the diseased patches more exactly by means of a light
thrown in from a reflector before treating them with
the brush. Gschirhakl described a similar apparatus
for endoscopic therapeutic purposes a few years ago.

We may employ any of the various solutions used
in blennorrhœa of the eyelids. Solutions of argentic nitrate in distilled water can be especially recommended ;
such as—

> ℞ Argent. nitrat .... 1·00
> Aquæ destillat............... .... 30·00
> Ft. sol.

---

* Where the urethra is very sensitive, it is well to introduce a
large steel sound (No. 25 or 26 French) just before using the brush
apparatus. The urethra is thus rendered anæsthetic, and the introduction of the smaller instrument and the application of the silver
nitrate solution now scarcely felt.—W. B. P.

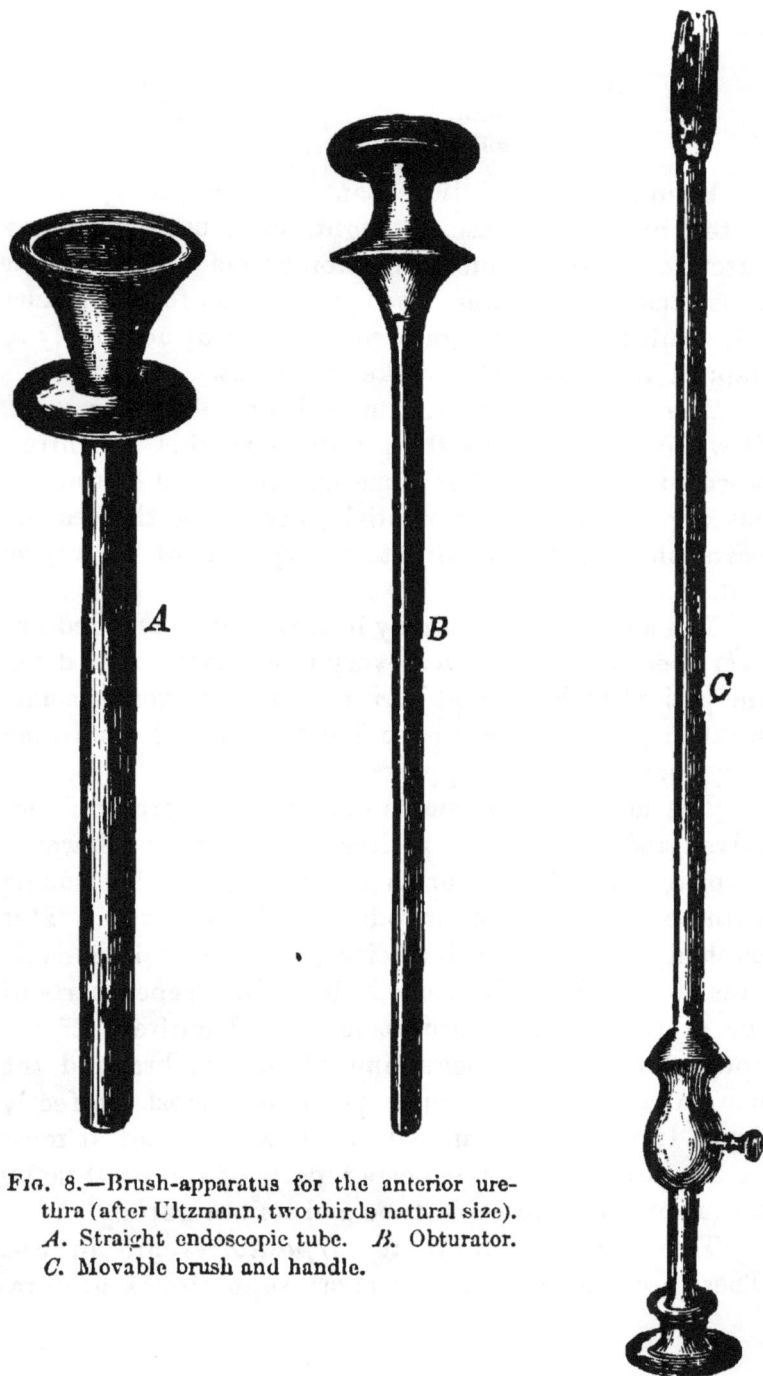

Fig. 8.—Brush-apparatus for the anterior ure-
thra (after Ultzmann, two thirds natural size).
A. Straight endoscopic tube. B. Obturator.
C. Movable brush and handle.

℞ Argent. nitrat.................. 1·00
    Aquæ destil................... 20·00
    Ft. sol.

Even stronger solutions can be employed up to one in ten, but in that case we ought not to use it over any extent of urethral mucous membrane, since great pain is thus caused ; and later, when the scab so formed comes off, urethral hæmorrhage occurs—an event not likely to happen after using the weaker solutions.

The solution most often used is one of five per cent. If we touch up the urethral walls with this, and afterward examine the alterations caused, we find the diseased places colored a whitish gray, while the healthy parts show only the slightest suspicion of a grayish tint.

The anterior urethra may in this way be brushed out every second day, or even every day. After it is done, the patient feels a slight burning, which increases somewhat in intensity during the first few minutes, and then very soon entirely disappears.

The use of medicaments in semi-solid form includes salves and urethral suppositories. Formerly it was a favorite method to cover an elastic olive-headed bougie with precipitate salve, introduce, and leave it in. This method is altogether inefficient, since it is impossible thus to introduce the remedy into the deeper parts of the urethra. The ointment is rubbed entirely off the bougie at the very beginning of the urethra, and the bougie reaches the deeper portions almost perfectly dry. If we are determined to use an ointment, it must be introduced into the deeper parts of the urethra by means of the endoscopic tube and its obturator.

Two kinds of urethral *suppositories* are in use. These are the long and the short suppositoria urethra-lia, which consist of the required medicament diffused through cocoa-butter or gelatine. They are in the form of narrow rods. The long urethral suppositories are generally prescribed to be used by the patient himself, who can introduce them into his urethra without the aid of the surgeon. The short urethral suppositories, on the contrary, are narrow rods two centimetres long, designed exclusively for the deeper parts of the urethra, and which can be introduced only by means of the straight tube and obturator. The long suppositories may be made of either cocoa-butter or gelatine, but cocoa-butter alone must be employed for the principal mass of the short suppositories. Gelatine has not the desirable firmness for such small affairs as these.

The amount of the medicament used in the suppository is the same, whether it be long or short, the difference being only in more or less of the excipient. The suppositories mostly used in chronic gonorrhœa are such as contain alum, zinc, copper, tannin, and the like. The most practical are the following :

    ℞ Alum. crud...................... 1·00
        Ol. theobrom..................... q. s.
    M. Ft. suppos. long. or brev. no. v.

    ℞ Tannin. pur................. 0·30–0·50
        Ol. theobrom..................... q. s.
        Ft. suppos. no. v.

    ℞ Zinc. sulphat................ 0·15–0·30
        Ol. theobrom..................... q. s.
    M. Ft. suppos. no. v.

One of these small affairs is introduced each day. The short suppositories are introduced into the bulbous urethra with the patient in the recumbent posture.

The patient either lies still after the introduction of the suppository, or, if he must go about, turns the penis up against the abdomen, compressing it beneath the body-band of his suspensory bandage, so as to prevent the melted suppository from escaping. The suppository ought to be retained in the urethra at least half an hour, and the patient should not urinate until at least this much time has elapsed.

Medicaments may be further used either in the form of powder or paste. Powders consist usually of alum, or of tannin, rubbed up with sugar, and blown into the urethra through the straight tube. Pastes which consist of the same medicaments mixed with some gelatine and gum are placed upon wax bougies, and these, as soon as dry, are introduced into the urethra. The medicinal bougies of Hochsinger are wax bougies prepared in this way, and contain zinc or tannin, alum or copper, or even argentic nitrate. These bougies are allowed to remain in the urethra until the paste has been melted off by the warmth of the urethra, and applied itself to the mucous membrane. Then the bougie is taken out of the urethra and the patient asked not to urinate before half an hour has elapsed. These bougies, as well as the urethral suppositories, may be used, according to circumstances, every day or every second day. In this place we must remark that there are varieties of urethritis which resist all these modes of local treatment. The persistence of such cases is usually due to some dyscrasia, and in that event constitutional, general treatment must be carried on at the same time. If syphilis be suspected, initiate an antisyphilitic treatment at once ; if tuberculosis, we should prescribe change of climate, or a sojourn in the country, and especially advise spending the winter in a warm southern clime. In these ways we often obtain the most gratifying results.

b. The Therapeutics of Suppuration from the Urethra

*when situated behind the Compressor Urethræ Muscle ;
Therapeutics of Catarrh of the Neck of the Bladder.*—
Catarrh of the neck of the bladder is sometimes acute,
again chronic. Acute catarrh of this part is usually a
consequence of gonorrhœa, and in this case we ought
never to employ instrumental interference. It is only
when retention of the urine sets in that we may introduce a soft elastic catheter, empty the bladder, and
wash it out with an antiseptic or narcotic solution. The
therapeutics of acute catarrh of the neck of the bladder
must be purely a regulation of the diet, and some internal medication. Since an acute catarrh of the neck
of the bladder is always accompanied by a catarrh of
the bladder itself, the treatment of this condition will
be considered along with that of the bladder in the
next chapter.

Chronic catarrh of the neck of the bladder is usually
the remnant of an obstinate gonorrhœal process which
has lasted for some time in the posterior urethra. Chronic catarrh of the neck of the bladder is one of those diseases which often resist even the most energetic local
treatment. We, moreover, find this catarrh in individuals who have practiced masturbation for some time,
led a very dissipated life, and been much given to sexual excesses. Besides, this catarrh occurs primarily in
beginning tuberculosis of the prostate, or of some other
portion of the genito-urinary apparatus. Chronic catarrh of the neck of the bladder can, as a rule, be cured
only by a rational local treatment. This consists in
depositing the medicament in the prostatic portion by
means of an appropriate instrument, or, if the medicament be used in a fluid form, it should be allowed to flow
through the prostatic portion itself. The treatment by
sounds is less fitted for catarrh of the neck of the blad-

der (i. e., the prostatic portion). Very often it causes an acute aggravation of the catarrh, with or without swelling of the prostate itself.

Local medication, by means of solutions, is carried out in different ways, according as dilute solutions of astringents are used or concentrated fluids for cauterizing. For dilute solutions irrigation by means of a *short* catheter is the best method. It is certainly sometimes possible to force a fluid into the bladder along the urethra without the use of a catheter. But this procedure is often very painful where the urethra is sensitive, and there is, at the same time, spasmodic contraction of the compressor urethræ muscle.

A much more convenient way is to overcome the spasmodically contracted compressor urethræ muscle by inserting a short catheter, and then proceed with the irrigation. By this method we always succeed perfectly, and with no pain worth mentioning. I described in detail an irrigating catheter of this sort some time since (see "Neuroses of the Genito-urinary Apparatus," Ultzmann, "Wiener Klinik," 1879). This consists of silver, is 16 cm. long, and has a diameter of Charrière Nos. 14–16. The vesical end has the usual medium curve of the metallic catheter, is smoothly rounded off, and either perforated with holes, like a sieve, or, as I have more recently ordered, provided with four longitudinal slits, 1 cm. long and 2–3 mm. wide each. The slits are to be preferred, since such a catheter is much easier to cleanse. The extra-vesical end has a disk of hard rubber on which is a mark to show in which direction the beak of the instrument points. A soft-rubber tube, 20 cm. long, is, in addition, attached to this extra-vesical end, that the irrigating syringe may be more easily connected.

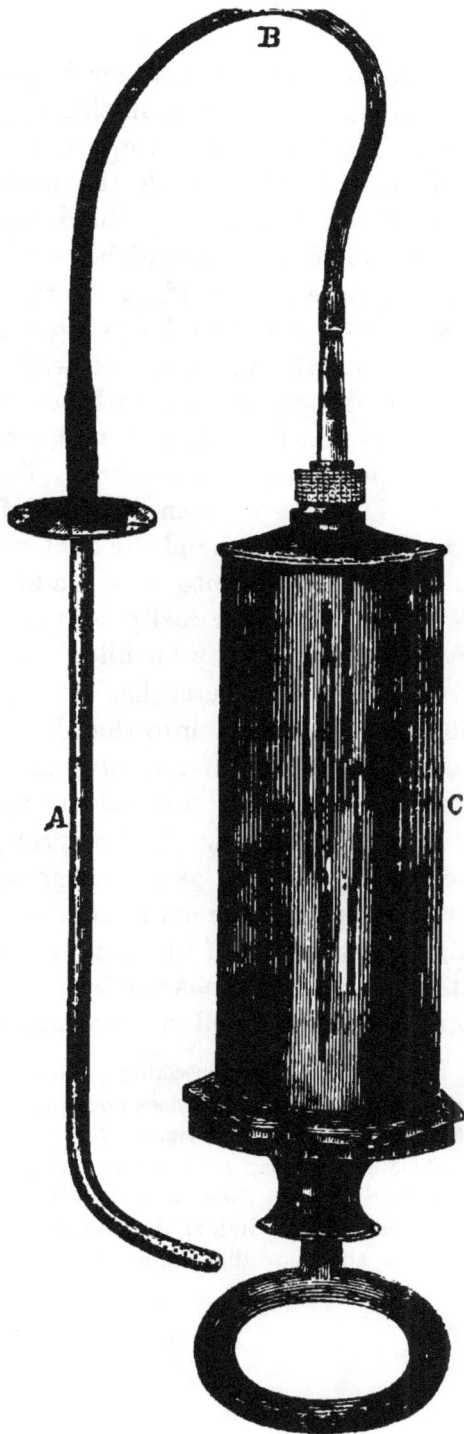

FIG. 9.—Irrigation catheter for the neck of the bladder (according to Ultzmann). *A*. Irrigation catheter. *B*. The connecting soft-rubber tube. *C*. The syringe.

Irrigation of the neck of the bladder is performed in the following manner by means of this apparatus : The patient lies on his back. A syringe holding 100–200 grammes of fluid is filled with the medicament and connected with the catheter by the India-rubber tube. The air is expelled from the catheter ; both the latter and the syringe are then held in the right hand, and the catheter is now introduced *lege artis* into the membranous portion of the urethra. If with the patient in the horizontal posture the catheter be tilted 30° beyond the perpendicular, its end will usually be in the membranous portion of the urethra. The catheter, thus introduced, is then taken in the left hand, firmly held, and with the right hand the contents of the syringe * are caused to flow into the bladder under gentle pressure. If the bladder easily contains a large · quantity of fluid, the syringe may be filled a second or even a third time. If the catheter has been correctly placed, the fluid can be injected into the bladder without any resistance being felt ; for the internal sphincter is such a weak muscle that it is not able to withstand the onward pressure of the fluid. If the syringe be removed from the catheter as soon as the irrigation is finished, no fluid will flow from the catheter if it has been properly placed, since the end of the catheter is not in the bladder but in the membranous urethra.

On the other hand, if the end of the catheter is in

---

* It is next to impossible to avoid injecting a little air into the bladder if we use a *syringe*. While this does no apparent harm in catarrh of the neck of the bladder, it is disagreeable to some patients. It can be entirely obviated by using a "fountain syringe" instead ; i. e., an irrigator, hung on the wall three or four feet above the patient. Let the warm fluid appear first at the end of the catheter, then introduce the latter, and allow the solution to slowly flow in.— W. B. P.

front of the compressor urethræ muscle (that is to say, in the bulb), the fluid can not be forced into the neck of the bladder, and will, in this latter case, flow out along the urethra by the side of the catheter, and thus escape from the meatus. Immediately after the irrigation, we have the patient empty his bladder completely.

It is only practicable and advisable to irrigate the neck of the bladder in this way, when we have to do with an entirely sufficient bladder—that is, when this latter can perfectly empty itself to the very last drop. If this is *not* the case, the irrigation by means of the short catheter had better be omitted, for, after the injection has taken place, an insufficient bladder could not perfectly empty itself of its contents, and necessarily a certain amount of fluid would remain behind, causing either a painful, constant desire to urinate, or other disturbances.

For an insufficient bladder, irrigation by means of an ordinary elastic catheter is much better. This is performed in the following way : The patient stands. He is caused to empty his bladder as much as he can spontaneously, then an elastic catheter is passed into the bladder. One of Mercier's catheters *coudé* is best, with two lateral openings. The amount of urine which can pass off now by the catheter is an index of the amount of the insufficiency of the bladder. When this latter is empty, draw out the catheter a little, so that the eyes of the instrument come to lie in the neck of the bladder ; this is now irrigated by means of a good hand-syringe. When a proper amount of fluid has been thus employed, push the catheter back again into the bladder, and empty it entirely. An ordinary washing-out of the bladder by a catheter is no substitute for this procedure ; since the eyes of the catheter lie in the bladder-cavity, the neck of the bladder is stopped up by the catheter, and thus the bladder itself is indeed washed out, but the neck of the bladder—the pars prostatica—not at all.

Many varieties of astringent aqueous solutions may be employed in irrigation of the neck of the bladder. The following can be recommended :

R Acid. carbolic................. .... 1·00
 Aquæ destillat.............................. 500·00
  Ft. sol.

R Alum. crud., zinci sulphat., acid. carbolic... āā 0·50–1·00
 Aquæ destillat.............................. 500·00
  Ft. sol.

R Potass. permanganat...................... 0·10–0·50
 Aquæ destillat.............................. 500·00
  Ft. sol.

R Argent. nitrat............................ 0·20–1·00
 Aquæ destillat........ :.................... 500·00
  Ft. sol.

It is best to warm these solutions before injecting.

If it is desired to treat the neck of the bladder with concentrated solutions to affect the part in a cauterizing rather than in an astringent manner, and confine the action to the neck of the bladder, the prostate alone, and leave the bladder itself untouched, in such a case we use a drop-catheter. Such an instrument I have previously described in connection with the local treatment of pollutions and spermatorrhœa. It consists of a short, catheter-like instrument of pure silver, having very thick walls, and a capillary calibre. It is of the same form, length, and (entire) diameter of the irrigation-catheter just described. The extra-vesical end is provided with a hard-rubber termination, into which a hypodermic syringe fits exactly. The capillary lumen of the catheter ends in the well-rounded tip. The instrument is so made that the entire capillary bore contains exactly as much as two divisions on the hypodermic syringe. If, now, we wish to inject one drop, we take up three into the syringe before attaching it to the catheter. If we would inject two drops, we fill the syringe with four drops, etc.

We employ the instrument in the following manner : The patient lies in the recumbent posture. We fill the hypodermic syringe with three to four drops of the solution, adapt it to the capillary-bore catheter, and then introduce the instrument *lege artis* into the pars prostatica, whereupon we force out the contents of the small syringe by gentle pressure with one finger. With the patient in the horizontal position, the end of the instrument will usually be in the prostatic portion, when the long axis of the catheter is 45° from the vertical. Should there be any uncertainty as to the position of the end of the instrument in the urethra, the fore-finger of the left hand in the rectum will always give us the desired information without further trouble. Immediately after cauterizing in this way, burning in the urethra sets in, and a few mo-

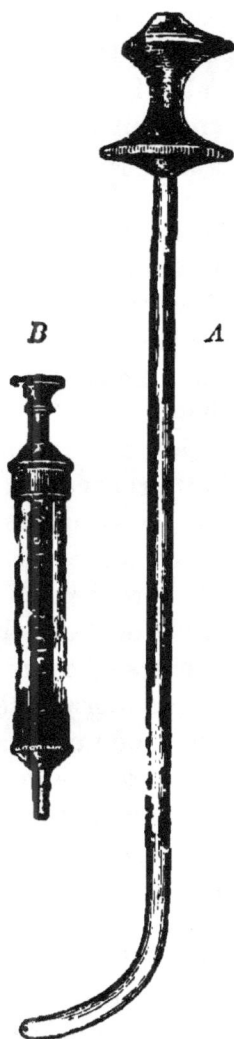

Fig. 10.—Urethral injector (according to Ultzmann). *A.* Capillary tubed catheter. *B.* Hypodermic syringe. (Two thirds natural size.)

ments later strong vesical tenesmus. Thus it is well for patients to remain rather quiet just after this procedure, and, when possible, to lie still for a little time until the tenesmus passes away.

Solutions of silver nitrate are usually employed, and of exactly the same strength as has already been given for brushing out the urethra with the brush apparatus. A five-per-cent solution is the one usually employed. If we wish to act more energetically, we may choose a ten-per-cent solution of argentic nitrate. In the latter case the cauterization is always followed by a hæmorrhage, sometimes very slight, sometimes considerable, which is not the case with the five-per-cent solution. With the weaker solution we may cauterize every second day, or even every day; while with the stronger solution we should cauterize but once, or at most twice, a week. This procedure is especially adapted for those slight catarrhs of the neck of the bladder, such as are apt to occur in cases of onanism, and after sexual excesses combined with abnormal seminal emissions. This is a most excellent mode of treatment in spermatorrhœa. And yet it is also of the greatest service in gonorrhœal catarrh.

Among the various methods of treating diseases by means of small suppositories, that of Dittels with the *porte-remède* is the best known and the most practical. By means of this *porte-remède*—a catheter-shaped, curved instrument, provided with an obturator—a small suppository is deposited in the prostatic portion. If we wish the suppository to act merely as an astringent on the prostatic portion of the urethra, we choose one of the suppositoria brevia previously described. If we desire to affect this region more actively, we choose a suppository made from cocoa-butter and argentic nitrate as follows :

℞ Argenti nitrat........................ 0·10
    Ol. theobromæ...................... q. s.
M. Ft. suppositor. urethral. brevia no. quinque.

If it is thought best to begin carefully, choose at
first only half a suppository. The caustic action of this
procedure is an extensive one, and not infrequently
followed by some bleeding from the urethra, lasting
several days. It is well if the patient keep in bed
meanwhile, although this is not absolutely necessary.

In catarrhs of the neck of the bladder, such as usually precede tuberculosis of the genito-urinary apparatus, we may use suppositoria brevia made of iodoform
and cocoa-butter, introducing them into the pars prostatica. These lessen somewhat the annoying tenesmus
of the bladder. They are prescribed in the following
manner :

℞ Iodoform. pur.,
    Ol. theobrom.................... āā q. s.
    Ft. suppos. urethral. brev. no. sex.

Finally, we can cauterize the pars prostatica with
lapis in substance, either with Lallemand's *porte-caus-
tique,* or, much better, by the use of the endoscope
under the control of the eye.

*c. Therapeutics of Suppuration of the Bladder it-
self; Therapeutics of Cystitis per se.*—Since every bladder-catarrh has its exciting cause, we can treat any
given case with success only by seeking out and trying
to remove the *causa movens.* There is no such thing
as a cystitis caused by "chill" or "catching cold."
Therefore we have no right to satisfy ourselves with
this as a convenient etiological factor. We ought all
the more to seek for the true cause by the most careful
kind of instrumental examination of the patient, and
not rest until we have found it.

The treatment of cystitis is very different, according as it is *acute* or *chronic*. Acute cystitis is treated according to the symptoms, and entirely by restricted diet and internal medication, while, in the treatment of chronic cystitis, local instrumental interference plays the principal part. *Acute catarrh of the bladder* is usually propagated from some neighboring section of the urinary tract, yet it may originate in the bladder itself. Chronic cystitis, on the contrary, usually has some parenchymatous or other organic change of the bladder itself or of the prostate behind it, or else there is something abnormal about the contents of the bladder. It is not to be overlooked that tuberculosis causes very obstinate forms of cystitis. From this simple remark alone we see that, in searching for the causes of a cystitis, we must often look for other etiological factors than bacteria, which are often wrongly accused of being the primary cause of the trouble, although they certainly are always present where ammoniacal fermentation is going on in urine.

In *acute catarrh of the bladder* (*acute cystitis*) rest in bed is indicated ; at least, the patient should remain in his room lying down. If there is high fever, rest in bed is absolutely necessary, since otherwise severe complications may occur, such as pyelitis and pyelo-nephritis, retention of urine, and hæmaturia. For the fever, we prescribe quinine, and if chills lasting some time take place, we order warm aromatic drinks, such as linden-flower tea or weak, ordinary table tea. For the pain over the region of the bladder, we order cataplasms made with decoctions of aromatic herbs, or linseed fomentations. If pain is present in the perinæum or rectum, a few leeches applied to the perinæum, or around the anus, will give a good deal of relief. For the troublesome, frequent, and painful desire to urinate, we prescribe narcotics, either internally or as suppositories. When given by the mouth, we can recommend morphia combined with lupulin, or sodium bicarbonate. Suppositories of morphia are far preferable to those of extractum belladonnæ. Extract of belladonna is a preparation of very uncertain (and unequal) action. While in one case we get next to no effect, a preparation from another apothecary may cause spastic contraction of the neck of the bladder, and thus only increase the strangury. The extract of hyoscyamus is still less reliable. The best way to prescribe these drugs is :

R   Lupulin. pur........................... 1·00
     Morphiæ muriat......................... 0·05
     Sacch. alb............................. 3·00
M. Ft. pulv. in dos. octo. Take 3–5 powders daily.

R   Morphiæ muriatic......... .... 0·10
     Ol. theobrom........................... 12·00
M. Ft. suppositor. no. sex. Use 2–3 daily.

If the suppositories are badly borne, we may use in their place small mucilaginous enemas, with ten to fifteen drops of tinctura opii two to three times daily. Milk, almond emulsion, and water are the best drinks. The decoctions of hemp and flaxseed, so often prescribed in acute cystitis, have no advantage over the above-mentioned drinks, and frequently nauseate. Mineral water or diuretic teas are *not* indicated as long as the urination remains very painful and quite frequent. We may only begin gradually with effervescent waters added to warm milk, when the painful tenesmus begins to lessen, and only when the pain has entirely disappeared use the undiluted mineral water. The same is true of the astringent and diuretic teas, such as folia uva ursæ, marrubium album, and chenopodium ambrosioides.

Warm sitz-baths, two to three times daily, or full baths, make urination considerably easier. If retention of urine set in, so that the bladder may be felt above the symphysis as a sensitive tumor, a soft catheter of vulcanized caoutchouc may be carefully introduced into the bladder, and the latter washed out, either with a one-tenth-per-cent lukewarm solution of carbolic acid, or, if much pain be present, with about three hundred grammes of tepid water, to which thirty drops of tinctura opii have been added. Stiff elastic or metallic catheters are less to be recommended, since they may easily injure the softened tissues of the neck of the bladder. The treatment of chronic cystitis ought to be always local, as distinguished from the acute form. Before we initiate local treatment we ought always to examine into the condition of the prostate as well as of the bladder with a steel sound or catheter. If the catarrh is kept up by a stone or a stricture, of course these must be disposed of first of all. If partial retention of urine, or inability of the bladder to entirely empty itself, be the prime factor in causing catarrh and ammoniacal fermentation, as is apt to be the case in hypertrophy of the prostate or paresis of the bladder, we must, above all else, secure complete and regular evacuation of the bladder by means of catheterism. We can employ medicated injections at the same time that we use this latter treatment. The local treatment of the vesical mucous membrane is always in the form of watery solutions which are injected into the bladder after this latter has been emptied by means of the catheter. The injection, as well as the washing out of the bladder, is best done in the following way : The patient stands ; after the bladder has been emptied in the natural way by the patient, an elastic catheter or a catheter *coudé* is passed into the bladder, and this latter *entirely* emptied. Now an injection is given by means of a syringe holding about one hundred grammes, employing moderate pressure. The first and the second syringeful should be allowed to flow out immediately after injecting, while the third syringeful may be retained a little longer. Or we may do in this way : First wash out the bladder with tepid water, by means of the syringe, until the washings come out clear and transparent ; and not until then inject the medicated solution, holding the catheter closed by the fingers for one or two minutes, after which time we allow the contents to flow away. This is the most practical way to treat a bladder locally. If the patient is weak or feverish, and can not stand, the washing is carried out in quite the same way with the patient in the recumbent or half-sitting posture. However, if he can stand, it is always preferable, since the sediment can be best got out of the bladder in this way. In former times washing out the bladder with a metallic catheter *à double courant* found much favor. This method is not good. In the first place, a metallic catheter irritates the bladder and the neck of the bladder much more than one made of soft vulcanized India-rubber ; and, in the second place, we always wash out the bladder when it is in a contracted state.

The water flows out of one opening in the catheter, and immediately back through the other opening. In this way but a small part of the surface of the mucous membrane of the bladder is rinsed by the water. The greater part of the catarrhal secretion, however, which sticks in the folds of the mucous membrane, or in the inter-trabecular spaces, can never be entirely washed out in this way. Any one can easily convince himself

that this is true by washing out a bladder affected with
purulent cystitis for a quarter of an hour with a double-
current catheter, and immediately afterward with the
hand-syringe and the soft-rubber catheter ; by this latter procedure a considerable amount of catarrhal secretion may be brought out, although, when washing with
the double-current catheter, the washings had become
quite clear. With each injection by means of the syringe
the bladder is distended, the sediment stirred up, and
thus more completely evacuated. Washing out the bladder by an irrigator is not a thoroughly good measure.
With an irrigator we usually allow the fluid to flow
quietly into the bladder until the patient has a feeling
of fullness and tension, whereupon the fluid is allowed
to flow out again. If the washing out is done in this
way in insufficient or paretic bladders they become only
more distended, and thus gradually more insufficient.

At the same time, the sediment is never so completely evacuated by this quiet flowing in and out of
the fluid as by the hand-syringe. In sensitive bladders all solutions ought to be injected warm. Cold
fluids cause a violent tenesmus at once, so that a contin-
uation of the washing out is impossible. On the contrary, in less sensitive bladders (as, for example, in
paresis), or in catarrhs complicated with hæmorrhages,
the injections ought to be made with cold solutions
only. Washing out the bladder is done with water
alone, or with water to which ten drops of tinct. opii
have been added to each 100 c. c.

Again, a $\frac{1}{8}$—$\frac{1}{4}$ per cent solution of carbolic acid
is well adapted for our purpose, or a 0·30 per cent solution of salicylic acid. Since this latter not seldom
causes a severe burning sensation, which is especially
apt to occur when the solution appears turbid from the

difficult solubility of the salicylic acid, we more often employ a 1–2 per cent solution of sodium salicylate instead. Stronger carbolic solutions of 1–1·5 per cent are only borne by less sensitive bladders. Also 3–5 per cent solutions of sodium chloride, or of sodium sulphate, or of borax, are especially to be recommended for washing out the bladder when the catarrhal secretion appears thick and ropy.

In many cases catheterism, in connection with washing out the bladder by one of the above methods, will of itself suffice to dispose of mucous or purulent catarrh of the bladder. However, in obstinate cases we must employ stronger astringents, such as potassium permanganate, 0·1–0·3 per cent solution ; of alum, a 1–5 per cent solution ; of zinc sulphate, a 0·3–2 per cent solution ; of zinc chloride, a 0·2–1 per cent solution ; and of silver nitrate, a 0·1 to a 0·5 per cent solution in a gradually increasing strength. If hæmorrhage of the bladder is present at the same time, besides the silver nitrate solution, a 0·5–2 per cent solution of ferrum sesquichloride is well adapted to the case, or a 0·5–2 per cent solution of tannin. In gangrenous, ichorous catarrh, we employ a 3–5 per cent solution of resorcin to obviate the fetor of the urine. The same purpose is served if we add 1–2 drops of amyl nitrite to every 100 grammes of water used for washing out the bladder. In order to bring the sediment present in *phosphaturia* partly into solution, we employ 0·2–0·1 per cent solutions of acid. hydrochloric. conc. In ammoniacal fermentation of the urine these solutions are used, alternating with the above astringent and antiseptic solutions. After severe gonorrhœal inflammations of the bladder, we sometimes find, besides a purulent cystitis, the capacity of the bladder very considerably lessened—so much so, indeed, that the patients may be obliged to urinate every five minutes, and prefer for convenience to wear a rubber urinal. In older patients the case is usually one of concentric hypertrophy of the bladder; in younger individuals, on the other hand, it is more apt to be a so-called "shrunken" or "cicatricial bladder." In these cases local treatment by means of medicated injections is of no avail if the capacity of the bladder can not be increased at the same time. This can be brought about by a gradual distention of the bladder (which has become so much smaller) with tepid water, by means of a syringe and a catheter. This method is not free from danger in old patients, since in hypertrophic bladders great diverticula are easily formed, and these might rupture. On the other hand, in the shrunken bladder from gonorrhœa, when occurring in individuals still young, its gradual disten-tion is followed by the very best results. If the puru-lent cystitis is but one of the manifestations of a gen-eral or local tuberculosis, or if the etiological factor is a compound one—as is often enough the case—local treatment will not be of much use unless the general con-dition is taken into consideration at the same time. It is a well-known fact that catarrhs of the bladder which defy all treatment can be made to disappear very soon if the patients spend a summer in the country, take an appropriate "cure"* at a spa, or a milk-cure, and

* A "cure" on the Continent is understood to mean a sojourn at a health resort, where the patients drink the waters of some mineral spring, bathe, and diet. In addition, massage and electricity are often employed. The habits of life can be exactly controlled, the surround-ings and occupation of the patient are altered, and by one or all of these influences the best results are often obtained in obstinate cases, when the treatment is judiciously conducted. A "cure," or course of treatment, usually takes six weeks.—W. B. P.

especially if they pass the winter in a southern climate, such as that of Italy. As an "after-cure," after a properly conducted local treatment, a "cure" at Carlsbad, Marienbad, Wildungen, or at one of the indifferent warm spas, as Gastein, Römerbad, Töplitz, and others, is of service. In cases of tuberculous cystitis, a "cure" at Gleichenberg or Roznau can be highly recommended.

d. Therapeutics of Suppuration of the Pelvis of the Kidney ; the Treatment of Pyelitis or Pyelo-nephritis.— The treatment of pyelitis is purely one of internal medication, since we can not get at the pelvis of the kidney with instruments for purposes of treatment. In extraordinary cases surgery has, to be sure, interfered repeatedly, yet until now the favorable results are small compared with the danger of the undertaking itself. Thus, kidneys have been repeatedly extirpated which were the seat of tumors, or movable, such as contained calculi, or had been injured, and in some cases the operation was followed by a good result. In the same way, incisions have been made into paranephritic abscesses, and into dilated pelves of the kidneys, in cases of pyonephrosis, sometimes to let out pus, sometimes to remove a calculus there situated.

Simply for the purpose of curing a purulent pyelitis no surgical interference has been attempted thus far. In those forms of pyelitis, such as set in during acute febrile processes, or in such as appear to result from setting back of urine in the bladder and are propagated from purulent catarrh in the latter, we have only to give our attention to the disease at the root of it all. If we can dispose of this, the pyelitis will usually vanish at the same time. *Acute pyelitis* demands for its treatment rest in bed. As long as fever is present, we must give quinine, and, if pain be present at the same time, morphia besides. For drink give milk, almond-milk, or water. If there is no great tenesmus present we may order a carbonated water, either by itself or given with milk. In *chronic pyelitis,* we order a milk cure and the systematic use of tepid full baths. To lessen suppuration, we prescribe tannin, tannate of quinine, and alum. To obviate any constipation that might be caused by these remedies, it is best to give with these a little ext. aloes aquos. The alum may be given in the form of alum-whey, of which half a litre must be drunk each day. These are good samples of prescriptions :

℞ Tannin. pur................................... 1·00
Sacchar. alb................................. 2·00
M. Ft. pulv. div. in dos. no. sex.
Take three powders daily.

℞ Quiniæ tannat...............................1·00
Sacchr. alb.................................2·00
Pulv. in dos. no. sex.
Give as in last prescription.

℞ Serum lactis clarif.......................500·00
Alum. crud. pulv......................... 3·00
D. S. Take during the day.

℞ Aq. calcis................................100·00
D. S. One to two tablespoonfuls to each glass of milk.

Balsam of copaiba and turpentine can sometimes be employed with good effect. They are both best given in gelatine capsules containing 0·20 each, of which six to twelve are to be taken daily. Turpentine is also often used as an inhalation, and can be employed cold or warm. Dittel recommends cold inhalations. For this purpose a teaspoonful of the purest turpentine is well shaken with about three hundred grammes of cold water, and a suitable quantity of this milky fluid inhaled through a kind of inhaler constructed like a "nargileh." The patient makes deep inspirations through the mouth-piece and tube of the instrument, and so introduces the turpentine into the system through the lungs. These inhalations are usually made several times daily, a few minutes at a time. If headache or dizziness is caused, the inhalations are to be suspended.

In *calculous pyelitis* we prescribe for the diathesis causing the calculus. Since uric acid and calcic oxalate are most frequent sources of renal calculi, the alkalies, and mineral waters containing them, are most appropriate. Thus, the waters of Carlsbad and Vichy have long been celebrated for their curative powers in lithiasis. However, all soda-springs have an excellent influence in this particular. To prevent formation of uric-acid calculi, lithia is especially recommended :

R Lithii carbonat.............................3·00
Ft. pulv. div. in dos. no. sex.
S. Three powders a day.

Several alkalies, given together, are also very appropriate for such cases. The following is an uncommonly good combination :

R Sodii phosphat...........................30·00
Sodii bicarbonat..........................60.00
Lithii carbonat...........................10·00
M. Ft. pulv.
Sig.: An even teaspoonful to be taken twice a day, dissolved in sweetened water.

That a "cure" at Carlsbad or Vichy is indicated in cases of calculous pyelitis, follows as a matter of course from what has been said above. If we suspect *pyelitis* or *pyelo-nephritis tuberculosa*, we ought, on the contrary, to bring into operation all those influences that show themselves efficacious in pulmonary phthisis ; thus a

5

cure at Gleichenberg, at Roznau, Meran, and in northern Italy, a corresponding dietetic treatment, and such medicine as the symptoms require. If in addition we get gastric troubles in chronic pyelo-nephritis, such as loss of appetite, .imperfect digestion, eructations, nausea, and vomiting, they are apt to be the symptoms of chronic uræmia ; and even if we can not hope for any permanent relief from any kind of treatment, it may be as well to state that the mineral acids taken internally can banish these troublesome symptoms for a time at least. We usually prescribe dilute hydrochloric acid in doses of ten to twenty drops, taken three times a day, after each meal ; or phosphoric acid, five grammes during the day in a sirup, and pure carbonated water in a siphon to drink.

THE END.

MEDICAL WORKS.

PYURIA; Or, PUS IN THE URINE, AND

ITS TREATMENT: Comprising the Diagnosis and Treatment of Acute and Chronic Urethritis, Prostatitis, Cystitis, and Pyelitis, with Especial Reference to their Local Treatment.

By Dr. ROBERT ULTZMANN,

Professor of Genito-Urinary Diseases in the Vienna Poliklinik.

Translated, by permission,

By Dr. WALTER B. PLATT, F. R. C. S. (Eng.).

12mo.

ANALYSIS OF THE URINE. With Special

Reference to the Diseases of the Genito-Urinary Organs.

By M. B. HOFFMAN,

Professor in the University of Gratz; and

R. ULTZMANN,

Docent in the University of Vienna.

Translated from the German edition under the special supervision of
Dr. Ultzmann,

By T. BARTON BRUNE, A. M., M. D.,

Resident Physician Maryland University Hospital; and

H. HOLBROOK CURTIS, Ph. B.

With Eight Lithographic Colored Plates from Ultzmann and Hoffman's Atlas, and from Photographs furnished by Dr. Ultzmann, which do not appear in the German edition or any other translation.

8vo.   Cloth, $2.00.

"It well deserves the reputation it has already obtained abroad. It is eminently practical, and adapted, as the authors claim, to the wants of the physician and student rather than of the medical chemist, although the latter can ill afford to dispense with it."—*Philadelphia Medical Times.*

MANUAL OF CHEMICAL EXAMINATION

OF THE URINE IN DISEASE. With Brief Directions for the Examination of the most Common Varieties of Urinary Calculi, and an Appendix containing a Diet-Table for Diabetics.

By AUSTIN FLINT, Jr., M. D.

Sixth edition, revised and corrected. 12mo.   Cloth, $1.00.

The chief aim of this little work is to enable the busy practitioner to make for himself, rapidly and easily, all ordinary examinations of Urine; to give him the benefit of the author's experience in eliminating little difficulties in the manipulations, and in reducing processes of analysis to the utmost simplicity that is consistent with accuracy.

New York: D. APPLETON & CO., 1, 3, & 5 Bond Street.

THE POPULAR SCIENCE MONTHLY.

CONDUCTED BY E. L. AND W. J. YOUMANS.

THE POPULAR SCIENCE MONTHLY will continue, as heretofore, to supply its readers with the results of the latest investigation and the most valuable thought in the various departments of scientific inquiry.

Leaving the dry and technical details of science, which are of chief concern to specialists, to the journals devoted to them, the MONTHLY deals with those more general and practical subjects which are of the greatest interest and importance to the public at large. In this work it has achieved a foremost position, and is now the acknowledged organ of progressive scientific ideas in this country.

The wide range of its discussions includes, among other topics:

The bearing of science upon education;

Questions relating to the prevention of disease and the improvement of sanitary conditions;

Subjects of domestic and social economy, including the introduction of better ways of living, and improved applications in the arts of every kind;

The phenomena and laws of the larger social organizations, with the new standard of ethics, based on scientific principles;

The subjects of personal and household hygiene, medicine, and architecture, as exemplified in the adaptation of public buildings and private houses to the wants of those who use them;

Agriculture and the improvement of food-products;

The study of man, with what appears from time to time in the departments of anthropology and archæology that may throw light upon the development of the race from its primitive conditions.

Whatever of real advance is made in chemistry, geography, astronomy, physiology, psychology, botany, zoölogy, paleontology, geology, or such other department as may have been the field of research, is recorded monthly.

Special attention is also called to the biographies, with portraits, of representative scientific men, in which are recorded their most marked achievements in science, and the general bearing of their work indicated and its value estimated.

The volumes begin with the May and November numbers, but subscriptions may begin at any time.

Terms: $5.00 per annum; single copy, 50 cents.

New York: D. APPLETON & CO., Publishers, 1, 3, & 5 Bond Street.

LECTURES UPON DISEASES OF THE

RECTUM AND THE SURGERY OF THE LOWER BOWEL.

Delivered at the Bellevue Hospital Medical College

By W. H. VAN BUREN, M. D.,

Late Professor of the Principles and Practice of Surgery in the Bellevue Hospital Medical College, etc., etc.

Second edition, revised and enlarged. 8vo, 412 pp., with 27 Illustrations and complete Index. Cloth, $3.00 ; sheep, $4.00.

The SCIENCE and ART of MIDWIFERY.

By WILLIAM THOMPSON LUSK, M. A., M. D.,

Professor of Obstetrics and Diseases of Women and Children in the Bellevue Hospital Medical College; Obstetric Surgeon to the Maternity and Emergency Hospitals; and Gynæcologist to the Bellevue Hospital.

Complete in one volume 8vo, with 226 Illustrations. Cloth, $5.00; sheep, $6.00.

"It contains one of the best expositions of the obstetric science and practice of the day with which we are acquainted. Throughout the work the author shows an intimate acquaintance with the literature of obstetrics, and gives evidence of large practical experience, great discrimination, and sound judgment. We heartily recommend the book as a full and clear exposition of obstetric science and safe guide to student and practitioner."—*London Lancet.*

"Professor Lusk's book presents the art of midwifery with all that modern science or earlier learning has contributed to it."—*Medical Record, New York.*

"This book bears evidence on every page of being the result of patient and laborious research and great personal experience, united and harmonized by the true critical or scientific spirit, and we are convinced that the book will raise the general standard of obstetric knowledge both in his own country and in this. Whether for the student obliged to learn the theoretical part of midwifery, or for the busy practitioner seeking aid in the face of practical difficulties, it is, in our opinion, the best modern work on midwifery in the English language."—*Dublin Journal of Medical Science.*

"Dr. Lusk's style is clear, generally concise, and he has succeeded in putting in less than seven hundred pages the best exposition in the English language of obstetric science and art. The book will prove invaluable alike to the student and the practitioner."—*American Practitioner.*

"Dr. Lusk's work is so comprehensive in design and so elaborate in execution that it must be recognized as having a status peculiarly its own among the text-books of midwifery in the English language."—*New York Medical Journal.*

"The work is, perhaps, better adapted to the wants of the student as a text-book, and to the practitioner as a work of reference, than any other one publication on the subject. It contains about all that is known of the *ars obstetrica*, and must add greatly to both the fame and fortune of the distinguished author."—*Medical Herald, Louisville.*

"Dr. Lusk's book is eminently viable. It can not fail to live and obtain the honor of a second, a third, and nobody can foretell how many editions. It is the mature product of great industry and acute observation. It is by far the most learned and most complete exposition of the science and art of obstetrics written in the English language. It is a book so rich in scientific and practical information, that nobody practicing obstetrics ought to deprive himself of the advantage he is sure to gain from a frequent recourse to its pages."—*American Journal of Obstetrics.*

"It is a pleasure to read such a book as that which Dr. Lusk has prepared; everything pertaining to the important subject of obstetrics is discussed in a masterly and captivating manner. We recommend the book as an excellent one, and feel confident that those who read it will be amply repaid."—*Obstetric Gazette, Cincinnati.*

"To consider the work in detail would merely involve us in a reiteration of the high opinion we have already expressed of it. What Spiegelberg has done for Germany, Lusk, imitating him but not copying him, has done for English readers, and we feel sure that in this country, as in America, the work will meet with a very extensive approval."—*Edinburgh Medical Journal.*

"The whole range of modern obstetrics is gone over in a most systematic manner, without indulging in the discussion of useless theories or controversies. The style is clear, concise, compact, and pleasing. The illustrations are abundant, excellently executed, remarkably accurate in outline and detail, and, to most of our American readers, entirely fresh."—*Cincinnati Lancet and Clinic.*

New York: D. APPLETON & CO., 1, 3, & 5 Bond Street.

EMERGENCIES, AND HOW TO TREAT

THEM. The Etiology, Pathology, and Treatment of Accidents, Diseases, and Cases of Poisoning, which demand Prompt Action. Designed for Students and Practitioners of Medicine.

By JOSEPH W. HOWE, M. D.,

Clinical Professor of Surgery in the Medical Department of the University of New York, etc., etc.

Third edition. 8vo, 265 pp. Cloth, $2.50.

"To the general practitioner in towns, villages, and in the country, where the aid and moral support of a consultation can not be availed of, this volume will be recognized as a valuable help. We commend it to the profession."—*Cincinnati Lancet and Observer.*

THE BREATH, AND THE DISEASES

WHICH GIVE IT A FETID ODOR. With Directions for Treatment.

By JOSEPH W. HOWE, M. D.,

Clinical Professor of Surgery in the Medical Department of the University of New York, etc.

Second edition, revised and corrected. 12mo, 108 pp. Cloth, $1.00.

"This little volume well deserves the attention of physicians, to whom we commend it most highly."—*Chicago Medical Journal.*
"To any one suffering from the affection, either in his own person or in that of his intimate acquaintances, we can commend this volume as containing all that is known concerning the subject, set forth in a pleasant style."—*Philadelphia Medical Times.*

A PRACTICAL TREATISE ON TUMORS

OF THE MAMMARY GLAND: embracing their Histology, Pathology, Diagnosis, and Treatment.

By SAMUEL W. GROSS, A. M., M. D.,

Surgeon to, and Lecturer on Clinical Surgery in, the Jefferson Medical College Hospital and the Philadelphia Hospital, etc.

In one handsome 8vo vol. of 246 pp., with 29 Illustrations. Cloth, $2.50.

"The work opportunely supplies a real want, and is the result of accurate work, and we heartily recommend it to our readers as well worthy of careful study."—*London Lancet.*

New York: D. APPLETON & CO., 1, 3, & 5 Bond Street.

A TREATISE ON THE PRACTICE OF

MEDICINE, for the Use of Students and Practitioners.

By ROBERTS BARTHOLOW, M. A., M. D., LL. D.,

Professor of Materia Medica and General Therapeutics in the Jefferson Medical College of Philadelphia; recently Professor of the Practice of Medicine and of Clinical Medicine in the Medical College of Ohio, in Cincinnati, etc., etc.

Fifth edition, revised and enlarged. 8vo. Cloth, $5.00; sheep or half russia, $6.00.

The same qualities and characteristics which have rendered the author's " Treatise on Materia Medica and Therapeutics " so acceptable are equally manifest in this. It is clear, condensed, and accurate. The whole work is brought up on a level with, and incorporates, the latest acquisitions of medical science, and may be depended on to contain the most recent information up to the date of publication.

" Probably the crowning feature of the work before us, and that which will make it a favorite with practitioners of medicine, is its admirable teaching on the treatment of disease. Dr. Bartholow has no sympathy with the modern school of therapeutical nihilists, but possesses a wholesome belief in the value and efficacy of remedies. He does not fail to indicate, however, that the power of remedies is limited, that specifics are few indeed, and that routine and reckless medication are dangerous. But throughout the entire treatise in connection with each malady are laid down well-defined methods and true principles of treatment. It may be said with justice that this part of the work rests upon thoroughly scientific and practical principles of therapeutics, and is executed in a masterly manner. No work on the practice of medicine with which we are acquainted will guide the practitioner in all the details of treatment so well as the one of which we are writing."—*American Practitioner.*

" The work as a whole is peculiar, in that it is stamped with the individuality of its author. The reader is made to feel that the experience upon which this work is based is real, that the statements of the writer are founded on firm convictions, and that throughout the conclusions are eminently sound. It is not an elaborate treatise, neither is it a manual, but halfway between: it may be considered a thoroughly useful, trustworthy, and practical guide for the general practitioner."—*Medical Record.*

" It may be said of so small a book on so large a subject, that it can be only a sort of compendium or *vade mecum*. But this criticism will not be just. For, while the author is master in the art of condensation, it will be found that no essential points have been omitted. Mention is made at least of every unequivocal symptom in the narration of the signs of disease, and characteristic symptoms are held well up in the foreground in every case."—*Cincinnati Lancet and Clinic.*

" Dr. Bartholow is known to be a very clear and explicit writer, and in this work, which we take to be his special life-work, we are very sure his many friends and admirers will not be disappointed. We can not say more than this without attempting to follow up the details of the plan, which, of course, would be useless in a brief book-notice. We can only add that we feel confident the verdict of the profession will place Dr. Bartholow's 'Practice' among the standard text-books of the day."—*Cincinnati Obstetric Gazette.*

" The book is marked by an absence of all discussion of the latest, fine-spun theories of points in pathology; by the clearness with which points in diagnosis are stated ; by the conciseness and perspicuity of its sentences; by the abundance of the author's therapeutic resources ; and by the copiousness of its illustrations."—*Ohio Medical Recorder.*

New York: D. APPLETON & CO., 1, 3, & 5 Bond Street.

THE APPLIED ANATOMY OF THE

NERVOUS SYSTEM, being a Study of this Portion of the Human Body from a Stand-point of its General Interest and Practical Utility, designed for Use as a Text-Book and as a Work of Reference.

By AMBROSE L. RANNEY, A. M., M. D.,

Adjunct Professor of Anatomy and late Lecturer on the Diseases of the Genito-Urinary Organs and on Minor Surgery in the Medical Department of the University of the City of New York, etc., etc.

8vo. Profusely illustrated. Cloth, $4.00; sheep, $5.00.

"This is a useful book, and one of novel design. It is especially valuable as bringing together facts and inferences which aid greatly in forming correct diagnoses in nervous diseases."—*Boston Medical and Surgical Journal.*

"This is an excellent work, timely, practical, and well executed. It is safe to say that, besides Hammond's work, no book relating to the nervous system has hitherto been published in this country equal to the present volume, and nothing superior to it is accessible to the American practitioner."—*Medical Herald.*

"There are many books, to be sure, which contain here and there hints in this field of great value to the physician, but it is Dr. Ranney's merit to have collected these scattered items of interest, and to have woven them into an harmonious whole, thereby producing a work of wide scope and of correspondingly wide usefulness to the practicing physician.

"The book, it will be perceived, is of an eminently practical character, and, as such, is addressed to those who can not afford the time for the perusal of the larger text-books, and who must read as they run."—*New York Medical Journal.*

"Professors of anatomy in schools and colleges can not afford to be without it. We recommend the book to practitioners and students as well."—*Virginia Medical Monthly.*

"It is an admitted fact that the subject treated of in this work is one sufficiently obscure to the profession generally to make any work tending to elucidation most welcome.

"We earnestly recommend this work as one unusually worthy of study."—*Buffalo Medical and Surgical Journal.*

"A useful and attractive book, suited to the time."—*Louisville Medical News.*

"Dr. Ranney has firmly grasped the essential features of the results of the latest study of the nervous system. His work will do much toward popularizing this study in the profession.

"We are sure that all our readers will be quite as much pleased as ourselves by its careful study."—*Detroit Lancet.*

"Our impressions of this work are highly favorable as regards its practical value to students, as well to educated medical men."—*Pacific Medical and Surgical Journal.*

"The work shows great care in its preparation. We predict for it a large sale among the more progressive practitioners."—*Michigan Medical News.*

"We are acquainted with no recent work which deals with the subject so thoroughly as this; hence, it should commend itself to a large class of persons, not merely specialists, but those who aspire to keep posted in all important advances in the science and art of medicine."—*Maryland Medical Journal.*

"This work was originally addressed to medical under-graduates, but it will be equally interesting and valuable to medical practitioners who still acknowledge themselves to be students. It is to be hoped that their number is not small."—*New Orleans Medical and Surgical Journal.*

"We think the author has correctly estimated the necessity for such a volume, and we congratulate him upon the manner in which he has executed his task.

"As a companion volume to the recent works on the diseases of the nervous system, it is issued in good time."—*North Carolina Medical Journal.*

"Dr. Ranney has done his work well, and given accurate information in a simple, readable style."—*Philadelphia Medical Times.*

New York: D. APPLETON & CO., 1, 3, & 5 Bond Street.